SpringerBriefs in Electrical and Computer
Engineering

For further volumes:
http://www.springer.com/series/10059

Thomas Robertazzi

Basics of Computer Networking

 Springer

Thomas Robertazzi
Stony Brook University
Stony Brook
NY, USA
e-mail: thomas.robertazzi@gmail.com

ISSN 2191-8112   e-ISSN 2191-8120
ISBN 978-1-4614-2103-0   e-ISBN 978-1-4614-2104-7
DOI 10.1007/978-1-4614-2104-7
Springer New York Dordrecht Heidelberg London

Library of Congress Control Number: 2011941592

Springer is part of Springer Science+Business Media (www.springer.com)

To Marsha,
My Late Wife and Partner

Preface

Computer networking is a fascinating field that has interested many for quite a few years. The purpose of this brief book is to give a general, non-mathematical, introduction to the technology of networks. This includes discussions of types of communication, many networking standards, popular protocols, venues where networking is important such as data centers, cloud computing and grid computing and the most important civilian encryption algorithm, AES.

This brief book can be used in undergraduate and graduate networking courses in universities or by the individual engineer, computer scientist or information technology professional. In universities it can be used in conjunction with more mathematical modeling oriented texts.

I have learned a great deal about networking by teaching undergraduate and graduate courses on the topic at Stony Brook. I am grateful to Dantong Yu of Brookhaven National Laboratory for making me aware of many recent technological developments. Thanks are also due to Brett Kurzman, my editor at Springer, for supporting this brief book project. I would like to acknowledge the assistance in my regular duties at the university of my department's superb staff of Gail Giordano, Carolyn Huggins, Rachel Ingrassia and Debbie Kloppenburg. I would also like to thank Prad Mohanty and Tony Olivo for excellent computer support.

The validation of my writing efforts by my daughters Rachel and Deanna and my good friend Sandy Pike means a lot. Finally I dedicate this brief book to the memory of my late wife and partner, Marsha.

Stony Brook, NY, September 2011   Thomas Robertazzi

Contents

Chapter 1
Introduction to Networks

1.1 Introduction

There is something about technology that allows people and their computers to communicate with each other that makes networking a fascinating field, both technically and intellectually.

What is a network? It is a collection of computers (nodes) and transmission channels (links) that allow people to communicate over distances, large and small. A Bluetooth personal area network may simply connect your home PC with its peripherals. An undersea fiber optic cable may traverse an ocean. The Internet and telephone networks span the globe.

Networking in particular has been a child of the late twentieth century. The Internet has been developed over the past 40 years or so. The 1980's and 1990's saw the birth and growth of local area networks, SONET fiber networks and ATM backbones. The 1990's and the early years of the new century have seen the development and expansion of WDM fiber multiplexing. New wireless standards continue to appear. Cloud computing and data centers are increasingly becoming a foundation of today's networking/computing world.

The book's purpose is to give a concise overview of some major topics in networking. We now start with an introduction to the applied aspects of networking.

1.2 Achieving Connectivity

A variety of transmission methods, both wired and wireless, are available today to provide connectivity between computers, networks and people. Wired transmission media include coaxial cable, twisted pair wiring and fiber optics. Wireless technology includes microwave line of sight, satellites, cellular systems, ad hoc networks and wireless sensor networks. We now review these media and technologies.

T. Robertazzi, *Basics of Computer Networking*, SpringerBriefs in Electrical and Computer Engineering, DOI: 10.1007/978-1-4614-2104-7_1,
© The authors 2012

1.2.1 Coaxial Cable

This is the thick cable you may have in your house to connect your cable TV set up box to the outside wiring plant. This type of cable has been around for many years and is a mature technology. While still popular for cable TV systems today, it was also a popular choice for wiring local area networks in the 1980's. It was used in the wiring of the original 10 Mbps Ethernet.

A coaxial cable has four parts: a copper inner core, surrounded by insulating material, surrounded by a metallic outer conductor; finally surrounded by a plastic outer cover. Essentially in a coaxial cable, there are two wires (copper inner core and outer conductor) with one geometrically inside the other. This configuration reduces interference to/from the coaxial cable with respect to other nearby wires.

The bandwidth of a coaxial cable is on the order of 1 GHz. How many bits per second can it carry? Modulation is used to match a digital stream to the spectrum carrying ability of the cable. Depending on the efficiency of the modulation scheme used, 1 bps requires anywhere from 1/14 to 4 Hz. For short distances, a coaxial cable may use 8 bits/Hz or carry 8 Gbps.

There are also different types of coaxial cable. One with a 50 ohm termination is used for digital transmissions. One with a 75 ohm termination is used for analog transmissions or cable TV systems.

A word is in order on cable TV systems. Such networks are locally wired as tree networks with the root node called the head end. At the head end, programming is brought in by fiber or satellite. From the head end cables (and possibly fiber) radiate out to homes. Amplifiers may be placed in this network when distances are large.

For many years, cable TV companies were interested in providing two way service. While early limited trials were generally not successful (except for Video on Demand), more recently cable TV seems to have winners in broadband access to the Internet and in carrying telephone traffic.

1.2.2 Twisted Pair Wiring

Coaxial cable is generally no longer used for wiring local area networks. One type of replacement wiring has been twisted pair. Twisted pair wiring typically had been previously used to wire phones to the telephone network. A twisted pair consists of two wires twisted together over their length. The twisted geometry reduces electromagnetic leakage (i.e. cross talk) with nearby wires. Twisted pairs can run several kilometers without the need for amplifiers. The quality of a twisted pair (carrying capacity) depends on the number of twists per inch.

Around 1990, it became possible to send 10 Mbps (for Ethernet) over unshielded twisted pair (UTP). Higher speeds are also possible if the cable and connector parameters are carefully implemented.

One type of unshielded twisted pair is category 3 UTP. It consists of four pairs of twisted pair surrounded by a sheath. It has a bandwidth of 16 MHz. Many offices used to be wired with category 3 wiring.

Category 5 UTP has more twists per inch. Thus, it has a higher bandwidth (100 MHz). Newer standards include category 6 versions (250 MHz or more) and category 7 versions (600 MHz or more). Category 8 at 1200 MHz is under development (Wikipedia).

The fact that twisted pair is lighter and thinner than coaxial cable has speeded its widespread acceptance.

1.2.3 Fiber Optics

Fiber optic cable consists of a silicon glass core that conducts light, rather than electricity as in coaxial cables and twisted pair wiring. The core is surrounded by cladding and then a plastic jacket.

Fiber optic cables have the highest data carrying capacity of any wired medium. A typical fiber has a capacity of 50 Tbps (terabits per second or 50×10^{12} bits/s). In fact, this data rate for years has been much higher than the speed at which standard electronics could load the fiber. This mismatch between fiber and nodal electronics speed has been called the "electronic bottleneck". Decades ago the situation was reversed, links were slow and nodes were relatively fast. This paradigm shift has led to a redesign of protocols.

There are two major types of fiber: multi-mode and single mode. Pulse shapes are more accurately preserved in single mode fiber, lending to a higher potential data rate. However, the cost of multi-mode and single-mode fiber is comparable. The real difference in pricing is in the opto-electronics needed at each end of the fiber. One of the reasons multi-mode fibers have a lower performance is dispersion. Under dispersion, square digital pulses tend to spread out in time, thus lowering the potential data rate. Special pulse shapes (such as hyperbolic cosines) called solitons, that dispersion is minimized for, have been the subject of research.

Mechanical fiber connectors to connect two fibers can lose 10% of the light that the fiber carries. Fusing two ends of the fiber results in a smaller attenuation.

Fiber optic cables today span continents and are laid across the bottom of oceans between continents. They are also used by organizations to internally carry telephone, data and video traffic.

1.2.4 Microwave Line of Sight

Microwave radio energy travels largely in straight lines. Thus, some network operators construct networks of tall towers kilometers apart and place microwave antennas at different heights on each tower. While the advantage is that there is no need to dig trenches for cables, the expense of tower construction and maintenance must be taken into account.

1.2.5 Satellites

Arthur C. Clarke, the science fiction writer, made popular the concept of using satellites as communication relays in the late 1940's. Satellites are now extensively used for communication purposes. They fill certain technological niches very well: providing connectivity to mobile users, for large area broadcasts and for communications for areas with poor infrastructure. The two main communication satellite architectures are geostationary satellites and low earth orbit satellites (LEOS). Both are now discussed.

1.2.5.1 Geostationary Satellites

You may recall from a physics course that a satellite in a low orbit (hundreds of kilometers) around the equator seems to move against the sky. As its orbital altitude increases, its apparent movement slows. At a certain altitude of approximately 36,000 km, it appears to stay in one spot in the sky, over the equator, 24 h a day. In reality, the satellite is moving around the earth but at the same angular speed that the earth is rotating, giving the illusion that it is hovering in the sky.

This is very useful. For instance, a satellite TV service can install home antennas that simply point to the spot in the sky where the satellite is located. Alternatively, a geostationary satellite can broadcast a signal to a large area (its "footprint") 24 h a day.

By international agreement, geostationary satellites are placed 2° apart around the equator. Some locations are more economically preferable than others, depending on which regions of the earth are under the location.

A typical geostationary satellite will have several dozen transponders (relay amplifiers), each with a bandwidth of 80 MHz [65]. Such a satellite may weigh several thousand kilograms and consume several kilowatts using solar panels.

The number of microwave frequency bands used have increased over the years as the lower bands have become crowded and technology has improved. Frequency bands include L (1.5/1.6 GHz), S (1.9/2.2 GHz), C (4/6 GHz), Ku (11/14 GHz) and Ka (20/30 GHz) bands. Here the first number is the downlink band and the second number is the uplink band. The actual bandwidth of a signal may vary from about 15 MHz in the L band to several GHz in the Ka band [65].

It should be noted that extensive studies of satellite signal propagation under different weather and atmospheric conditions have been conducted. Excess power for overcoming rain attenuation is often budgeted above 11 GHz.

Fig. 1.1 Low earth orbit
satellites (LEOS) in polar
orbits

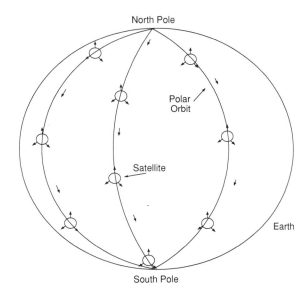

1.2.5.2 Low Earth Orbit Satellites

A more recent architecture is that of low earth orbit satellites. The most famous such system was Iridium from Motorola. It received its name because the original proposed 77 satellite network has the same number of satellites as the atomic number of the element Iridium. In fact, the actual system orbited had 66 satellites but the system name Iridium was kept.

The purpose of Iridium was to provide a global cell phone service. One would be able to use an Iridium phone anywhere in the world (even on the ocean or in the Artic). Unfortunately, after spending five billion dollars to deploy the system, talking on Iridium cost a dollar or more a minute while local terrestrial cell phone service was under 25 cents a minute. While an effort was made to appeal to business travelers, the system was not profitable and was sold and is now operated by a private company.

Technologically though, the Iridium system is interesting. There are 11 satellites in each of six polar orbits (passing over the North Pole, south to the South Pole and back up to the North Pole, see Fig. 1.1).

At any given time, several satellites are moving across the sky over any location on earth. Using several dozen spot beams, the system can support almost a quarter of a million conversations. Calls can be relayed from satellite to satellite.

It should be noted that when Iridium was hot, several competitors were proposed but not built. One used a "bent pipe" architecture where a call to a satellite would be beamed down from the same satellite to a ground station and then sent over the terrestrial phone network rather than being relayed from satellite to satellite. This was done in an effort to lower costs and simplify the design.

Fig. 1.2 Part of a cellular
network

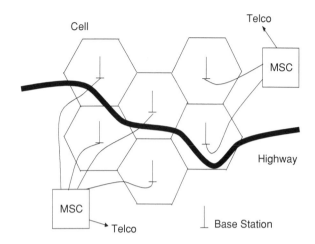

1.2.6 Cellular Systems

Starting around the early 1980's, cellular telephone systems which provide connectivity between mobile phones and the public switched telephone network were deployed. In such systems, signals go from/to a cell phone to/from a local "base station" antenna which is hard wired into the public switched telephone network. Figure 1.2 illustrates such a system. A geographic region such as a city or suburb is divided into geographic sub-regions called "cells".

Base stations are shown at the center of cells. Nearby base stations are wired into a switching computer (the mobile switching center or MSC) that provides a path to the telephone network.

A cell phone making a call connects to the nearest base station (i.e., the base station with the strongest signal). Base stations and cell phones measure and communicate received power levels. If one is driving and one approaches a new base station, its signal will at some point become stronger than that of the original base station one is connected to and the system will then perform a "handoff". In a handoff, connectivity is changed from one base station to an adjacent one. Handoffs are transparent, the talking user is not aware when one occurs.

Calls to a cell phone involve a paging like mechanism that activates (rings) the called user's phone.

The first cellular system was deployed in 1979 in Japan by NTT. The first US cellular system was AMPS (Advanced Mobile Phone System) from AT&T. It was first deployed in 1983. These were first generation analog systems. Second generation systems were digital. The most popular is the European originated GSM (Global System for Mobile), what has been installed over the world. Third and fourth generation cellular systems provide increased data rates for such applications as Internet browsing and picture transmission.

1.2.7 Ad Hoc Networks

Ad hoc networks [39, 45] are radio networks where (often mobile) nodes can come together, transparently form a network without any user interaction and maintain the network as long as the nodes are in range of each other and energy supplies last [34, 49]. In an ad hoc network messages hop from node to node to reach an ultimate destination. For this reason ad hoc networks used to be called multi-hop radio networks. In fact, because of the nonlinear dependence of energy on transmission distance, the use of several small hops uses much less energy than a single large hop, often by orders of magnitude.

Ad hoc network characteristics include multi-hop transmission, possibly mobility and possibly limited energy to power the network nodes. Applications include mobile networks, emergency networks, wireless sensor networks and ad hoc gatherings of people, as at a convention center.

Routing is an important issue for ad hoc networks. Two major categories of routing algorithms are topology-based routing and position-based routing. Topology-based routing uses information on current links to perform the routing. Position-based routing makes use of a knowledge of the geographic location of each node to route. The position information may be acquired from a service such as the Global Positioning System (GPS).

Topology-based algorithms may be further divided into proactive and reactive algorithms. Proactive algorithms use information on current paths as inputs to classical routing algorithms. However to keep this information current a large amount of control message traffic is needed, even if a path is unused. This overhead problem is exacerbated if there are many topology changes (say due to movement of the nodes).

On the other hand, reactive algorithms such as DSR, TORA and AODV maintain routes only for paths currently in use to keep the amount of information and control overhead more manageable. Still, more control traffic is generated if there are many topology changes.

Position-based routing does not require maintenance of routes, routing tables, or generation of large amounts of control traffic other than information regarding positions. "Geocasting" to a specific area can be simply implemented. A number of heuristics can be used in implementing position-based routing.

1.2.8 Wireless Sensor Networks

The integration of wireless, computer and sensor technology has the potential to make possible networks of miniature elements that can acquire sensor data and transmit the data to a human observer. Wireless sensor networks have received attention from researchers in universities, government and industries because of their promise to become a revolutionary technology and the technical challenges that must be overcome to make this a reality. It is assumed that such wireless sensor networks will use ad hoc radio networks to forward data in a multi-hop mode of operation.

Typical parameters for a wireless sensor unit (including computation and networking circuitry) include a size from 1 mm to 1 cm, a weight less than 100 g, cost less than one dollar and power consumption less than 100 μW [59]. By way of contrast, a wireless personal area network Bluetooth transceiver consumes more than a 1000 μW. A cubic millimeter wireless sensor can store, with battery technology, 1 Joule allowing a 10 μW energy consumption for 1 day [20]. Thus energy scavenging from light or vibration has been proposed. Note also that data rates are often relatively low for sensor data (100's bps–100 Kbps).

Naturally, with these parameters, minimizing energy usage in wireless sensor networks becomes important. While in some applications wireless sensor networks may be needed for a day or less, there are many applications where a continuous source of power is necessary. Moreover, communication is much more energy expensive than computation. Sending one bit for a distance of 100 m can take as much energy as processing 3,000 instructions on a micro-processor.

While military applications of wireless sensor networks are fairly obvious, there are many potential scientific and civilian applications of wireless sensor networks. Scientific applications include geophysical, environmental and planetary exploration. One can imagine wireless sensor networks being used to investigate volcanoes, measure weather, monitor beach pollution or record planetary surface conditions.

Biomedical applications include applications such as glucose level monitoring and retinal prosthesis [58]. Such applications are particularly demanding in terms of manufacturing sensors that can survive in and not affect the human body.

Sensors can be placed in machines (where vibration can sometimes supply energy) such as rotating machines, semiconductor processing chambers, robots and engines. Wireless sensors in engines could be used for pollution control telemetry.

Finally, among many potential applications, wireless sensors could be placed in homes and buildings for climate control. Note that wiring a single sensor in a building can cost several hundred dollars. Ultimately, wireless sensors could be embedded in building materials.

1.3 Multiplexing

Multiplexing involves sending multiple signals over a single medium. Thomas Edison invented a four to one telegraph multiplexer that allowed four telegraph signals to be sent over one wire. The major forms of multiplexing for networking today are frequency division multiplexing (FDM), time division multiplexing (TDM) and spread spectrum. Each is now reviewed.

1.3.1 Frequency Division Multiplexing (FDM)

Here a portion of spectrum (i.e. band of frequencies) is reserved for each channel (Fig. 1.3a). All channels are transmitted simultaneously but a tunable filter at the receiver only allows one channel at a time to be received. This is how AM, FM

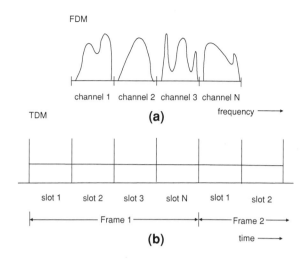

Fig. 1.3 **a** Frequency division multiplexing, **b** Time division multiplexing

and analog television signals are transmitted. Moreover, it is how distinct optical signals are transmitted over a fiber using wavelength division multiplexing (WDM) technology.

1.3.2 Time Division Multiplexing (TDM)

Time division multiplexing is a digital technology that, on a serial link, breaks time into equi-duration slots (Fig. 1.3b). A slot may hold a voice sample in a telephone system or a packet in a packet switching system. A frame consists of N slots. Frames, and thus slots, repeat. A telephone channel might use slot 14 of 24 slots in a frame during the length of a call, for instance.

Time division multiplexing is used in the second generation cellular system, GSM. It is also used in digital telephone switches. Such switches in fact use electronic devices called time slot interchangers that transfer voice samples from one slot to another to accomplish switching.

1.3.3 Frequency Hopping

Frequency hopping is one form of spread spectrum technology and is typically used on radio channels. The carrier (center) frequency of a transmission is pseudo-randomly hopped among a number of frequencies (Fig. 1.4a). The hopping is done in a deterministic, but random looking pattern that is known to both transmitter and receiver. If the hopping pattern is known only to the transmitter and receiver, one has

Fig. 1.4 a Frequency hopping spread spectrum, **b** Direct sequence spread spectrum

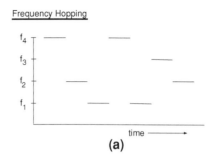

Table 1.1 XOR Truth Table

Key	Data	Output
0	0	0
0	1	1
1	0	1
1	1	0

good security. Frequency hopping also provides good interference rejection. Multiple transmissions can be multiplexed in the same local region if each uses a sufficiently different hopping pattern. Frequency hopping dates back to the era of World War II.

1.3.4 Direct Sequence Spread Spectrum

This alternative spread spectrum technology uses exclusive or (xor) gates as scramblers and de-scramblers (Fig. 1.4b). At the transmitter data is fed into one input of an xor gate and a pseudo-random key stream into the other input (Table. 1.1).

From the xor truth table, one can see that if the key bit is a zero, the output bit equals the data bit. If the key bit is a one, the output bit is the complement of the data bit (0 becomes 1, 1 becomes 0). This scrambling action is quite strong under the proper conditions. Unscrambling can be performed by an xor gate at the receiver. The transmitter and receiver must use the same (synchronized) key stream for this to work. Again, multiple transmissions can be multiplexed in a local region if the key streams used for each transmission are sufficiently different.

Fig. 1.5 a Circuit switching,
b Packet switching

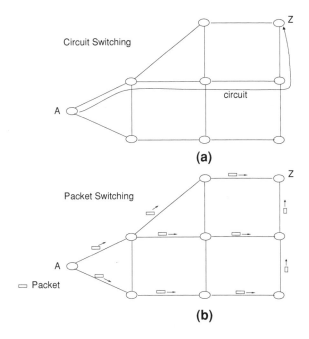

(a)

(b)

1.4 Circuit Switching Versus Packet Switching

Two major architectures for networking and telecommunications are circuit switching and packet switching. Circuit switching is the older technology, going back to the years following the invention of the telephone in the late 1800's. As illustrated in Fig. 1.5a, for a telephone network, when a call has to be made from node A to node Z, a physical path with appropriate resources called a "circuit" is established. Resources include link bandwidth and switching resources. Establishing a circuit requires some set-up time before actual communication commences. Even if one momentarily stops talking, the circuit is still in operation. When the call is finished, link and switching resources are released for use by other calls. If insufficient resources are available to set up a call, the call is said to be blocked.

Packet switching was created during the 1960's. A packet is a bundle of bits consisting of header bits and payload bits. The header contains the source and destination address, priority levels, error check bits and any other information that is needed. The payload is the actual information (data) to be transported. However, many packet switching systems have a maximum packet size. Thus, larger transmissions are split into many packets and the transmission is reconstituted at the receiver.

The diagram of Fig. 1.5b shows packets, possibly from the same transmission, taking multiple routes from node A to node Z. This is called datagram or connectionless oriented service. Packets may indeed take different routes in this type of service as nodal routing tables are updated periodically in the middle of a transmission.

A hybrid type of service is the use of "virtual circuits" or connection oriented service. Here packets belonging to the same transmission are forced to take the same serial path through the network. A virtual circuit has an identification number which is used at nodes to continue the circuit along its preset path. As in circuit switching, a virtual circuit needs to be set up prior to its use for communication. That is, entries need to be made in routing tables implementing the virtual circuit.

An advantage of virtual circuit usage is that packets arrive at the destination in the same order that they were sent. This avoids the need for buffers for reassembling transmissions (reassembly buffers) that are needed when packets arriving at the destination are not in order, as in datagram service. As we shall see, ATM, the high speed packet switching technology used in Internet backbones, uses virtual circuits.

Packet switching is advantageous when traffic is bursty (occurs at irregular intervals) and individual transmissions are short. It is a very efficient way of sharing network transmissions when there are many such transmissions. Circuit switching is not well suited for bursty and short transmissions. It is more efficacious when transmissions are relatively long (to minimize set up time overhead) and provide a constant traffic rate (to well utilize the dedicated circuit resource).

1.5 Layered Protocols

Protocols are the rules of operation of a network. A common way to engineer a complex system is to break it into more manageable and coherent components. Network protocols are often divided into layers in the layered protocol approach. Figure 1.6 illustrates the generic OSI (open systems interconnection) protocol stack. Proprietary protocols may have different names for the layers and/or a different layer organization but pretty much all networking protocols have the same functionality.

Transmissions in a layered architecture (see Fig. 1.6) move from the source's top layer (application), down the stack to the physical layer, through a physical channel in a network, to the destination's physical layer, up the destination stack to the destination application layer. Note that any communication between peer layers must move down one stack, across and up the receiver's stack. It should also be noted that if a transmission passes through an intermediate node, only some lower layers (e.g., network, data link and physical) may be used at the intermediate nodes.

It is interesting that a packet moving down the source's stack may have its header grow as each layer may append information to the layer. At the destination, each layer may remove information from the packet header, causing it to decrease in size as it moves up the stack.

In a particular implementation, some layers may be larger and more complex while others are relatively simple.

In the following, we briefly discuss each layer.

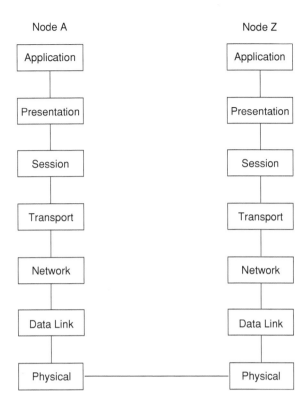

Fig. 1.6 OSI protocol stack for a communicating source and destination

1.5.1 Application Layer

Applications for networking include email, remote login, file transfer and the world-wide web. But an application may also be more specialized, such as distributed software to run a network of catalog company order depots.

1.5.2 Presentation Layer

This layer controls how information is formatted, such as on a screen (number of lines, number of characters across).

1.5.3 Session Layer

This layer is important for managing a session, as in remote logins. In other cases, this is not a concern.

1.5.4 Transport Layer

This layer can be thought of as an interface between the upper and lower layers. More importantly, it is designed to give the impression to the layers above that they are dealing with a reliable network, even though the layers below the transport layer may not be perfectly reliable. For this reason, some think of the transport layer as the most important layer.

1.5.5 Network Layer

The network layer manages multiple links. Its most important function is to do routing. Routing involves selecting the best path for a circuit or packet stream.

1.5.6 Data Link Layer

Whereas, the network layer manages multiple link functions, a data link protocol manages a single link. One of its potential functions is encryption, which can either be done on a link-by-link basis (i.e. at the data link layer) or on an end-to-end basis (i.e. at the transport layer) or both. End-to-end encryption is a more conservative choice as one is never sure what type of sub-network a transmission may pass thru and what its level of encryption, if any, is.

1.5.7 Physical Layer

The physical layer is concerned with the raw transmission of bits. Thus, it includes engineering physical transmission media, modulation and de-modulation and radio technology. Many communication engineers work on physical layer aspects of networks. Again, the physical layer of a protocol stack is the only layer that provides actual direct connectivity to peer layers.

Introductory texts on networking usually discuss layered protocols in detail.

Chapter 2
Ethernet

2.1 Introduction

Local area networks (LANs) are networks that cover a small area as in a department
in a company or university. In the early 1980s, the three major local area networks
were Ethernet (IEEE standard 802.3), Token Ring (802.5 and used extensively by
IBM) and Token Bus (802.4, intended for manufacturing plants). However, over the
years, Ethernet [65] has become the most popular local area network standard. While
maintaining a low cost, it has gone through six versions, most ten times faster than
the previous version (10 Mbps, 100 Mbps, 1 Gbps, 10 Gbps, 40 Gbps, 100 Gbps).

Ethernet was invented at the Xerox Palo Alto Research Center (PARC) by Metcalfe
and Boggs, circa [36]. It is similar in spirit to the earlier Aloha radio protocol, though
the scale is smaller. IEEE's 802.3 committee produced the first Ethernet standard.
Xerox never produced Ethernet commercially but other companies did.

In going from one Ethernet version to the next, the IEEE 802.3 committee sought
to make each version similar to the previous ones and to use existing technology.
In the following, we now discuss the various versions of Ethernet.

2.2 10 Mbps Ethernet

Back in the 1980s, Ethernet was originally wired using coaxial cable. As in Fig. 2.1a,
a coaxial cable was snaked through the floor or ceiling and computers attached to it
along its length. The coaxial cable acted as a private radio channel that each computer
would monitor. If a station had a packet to send, it would send it immediately if the
channel was idle. If the station sensed the channel to be busy, it would wait until the
channel was free. In all of this, only one transmission can be on the channel at one
time.

A problem occurs if two or more stations sense the channel to be idle at about the
same time and attempt to transmit simultaneously. The packets overlap in the cable

T. Robertazzi, *Basics of Computer Networking*, SpringerBriefs in Electrical
and Computer Engineering, DOI: 10.1007/978-1-4614-2104-7_2,
© The authors 2012

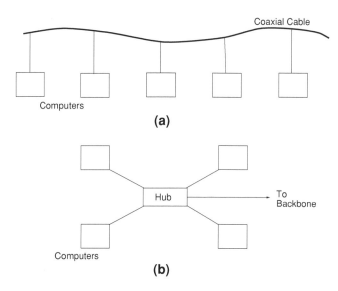

Fig. 2.1 Ethernet wiring using (**a**) coaxial cable and (**b**) hub topology

and are garbled. This is a collision. The stations involved, using analog electronics, can detect the collision, stop transmitting and reschedule their transmissions.

Thus, the price one pays for this completely decentralized access protocol is the presence of utilization lowering collisions. The protocol used goes by the name 1-persistent Carrier Sense Multiple Access with Collision Detection (CSMA/CD). The name is pretty much self-explanatory except that 1-persistent refers to the fact that a station with a packet to send attempts this on an idle channel with a probability of 1.0. In a CSMA/CD protocol, if the bit rate is 10 Mbps, the actual useful information transport can be significantly less because of collisions (or occasional idleness).

In the case of a collision, the rescheduling algorithm used is called Binary Exponential Backoff. Under this protocol, two or more stations experiencing a collision randomly reschedule over a time window with a default of 51 microseconds for a 500 m network. If a station becomes involved in a second collision, it doubles its window size and attempts again to randomly reschedule its transmission. Windows may be doubled in size up to ten times. Once a packet is successfully transmitted, the window size drops back to the default (smallest) value for that packet's station. Thus, this protocol at a station has no long-term memory regarding past transmissions.

Table 2.1 shows the fields in the 10 Mbps Ethernet frame. A frame is the name for a packet at the data link layer. The preamble is for communication receiver synchronization purposes. Addresses are either local (2 bytes) or global (6 bytes). Note that Ethernet addresses are different from IP addresses. Different amounts of data can be accommodated up to 1500 bytes. Transmissions longer than 1500 bytes of data must be segmented into multiple packets. The pad field is used to guarantee that the frame is at least 64 bytes in length (minimum frame size) if the frame would

Table 2.1 Ethernet frame format

Field	Length
Preamble	7 bytes
Frame delimiter	1 byte
Destination address	2 or 6 bytes
Source address	2 or 6 bytes
Data length	2 bytes
Data	up to 1,500 bytes
Pad	variable
CRC checksum	4 bytes

be less than 64 bytes in length. Finally the checksum is based on CRC error detecting coding.

A problem with digital receivers is that they require many 0 to 1 and 1 to 0 transitions to properly lock onto a signal. But long runs of 1's or 0's are not uncommon in data. To provide many transitions between logic levels, even if the data has a long run of one logic level, 10 Mbps Ethernet uses Manchester encoding.

Referring to Fig. 2.2, under Manchester encoding, if a logic 0 needs to be sent, a transition is made for 0 to 1 (low to high voltage) and if a logic 1 needs to be sent, the opposite transition is made for 1 to 0 (high to low voltage). The voltage level makes a return to its original level at the end of a bit as necessary. Note that the "signaling rate" is variable. That is, the number of transitions per second is twice the data rate for long runs of a logic level and is equal to the data rate if the logic level alternates. For this reason, Manchester encoding is said to have an efficiency of 50%. More modern signaling codes, such as 4B5B, achieve a higher efficiency (see Fast Ethernet).

During the 1980s, Ethernets were wired with linear coaxial cables. Today hubs are commonly used (Fig. 2.1b). These are boxes (some smaller than a cigar box) that computers tie into, in a star type wiring pattern, with the hub at the center of the star.

A hub may internally have multiple cards, each of which have multiple external Ethernet connections. A high speed (in the gigabits) proprietary bus interconnects the cards. Cards may mimic a CSMA/CD Ethernet with collisions (shared hub) or use buffers at each input (switched hub). In a switched hub, multiple packets may be received simultaneously without collisions, raising throughput.

The next table (Table 2.2) illustrates Ethernet wiring. In "10 Base5", the 10 stands for 10 Mbps and the 5 for the 500 m maximum size. Used in the early 1980s, 10 Base5 used vampire taps which would puncture the cable. Also at the time, 10 Base2 used T junctions and BNC connectors as wiring hardware. Today, 10 Base-T is the most common wiring solution for 10 Mbps Ethernet. Fiber optics, 10 Base-F, is only intended for runs between buildings, but a higher data rate protocol would probably be used today for this purpose.

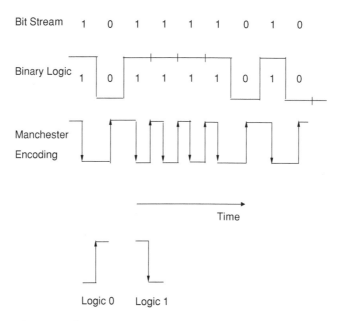

Fig. 2.2 Manchester encoding

Cable	Type	Maximum size
10Base5	Thick coax	500 m
10Base2	Thin coax	200 m
10Base-T	Twisted pair	100 m
10Base-F	Fiber optics	2 km

Table 2.2 Original Ethernet wiring

2.3 Fast Ethernet

As the original 10 Mbps Ethernet became popular and the years passed, traffic on Ethernet networks continued to grow. To maintain performance, network administrators were forced to segment Ethernet networks into smaller networks (each handling a smaller number of computers) connected by a spaghetti-like arrangement of repeaters, bridges and routers. In 1992, IEEE assigned the 802.3 committee the task of developing a faster local area network protocol.

The committee agreed on a 100 Mbps protocol that would incorporate as much of the existing Ethernet protocol/technology as possible to gain acceptance and so that they could move quickly. The resulting protocol, IEEE 802.3u, was called Fast Ethernet.

Fast Ethernet is only implemented with hubs, in a star topology (Fig. 2.1b). There are three major wiring options (Table 2.3).

The original Ethernet has a data rate of 10 Mbps and a maximum signaling rate of 20 MHz (recall that the Manchester encoding used was 50% efficient). Fast Ethernet

Cable	Type	Maximum size
100Base-T4	Twisted pair	100 m
100Base-TX	Twisted pair	100 m
100Base-FX	Fiber optics	2 km

Table 2.3 Fast Ethernet wiring

100 Base-T4 with its data rate of 100 Mbps has a signaling speed of 25 MHz, not 200 MHz. How is this accomplished?

Fast Ethernet 100 Base-T4 actually uses four twisted pairs per cable. Three twisted pairs carry signals from its hub to a PC. Each of the three twisted pairs uses ternary (not binary) signaling using 3 logic levels. Thus, one of $3 \times 3 \times 3 = 27$ symbols can be sent at once. Only 16 symbols are used though, which is equivalent to sending 4 bits at once. With 25 MHz clocking 25 MHz $\times$ 4 bits yields a data rate of 100 Mbps. The channel from the PC to hub operates at 33 MHz. For most PC applications, an asymmetrical connection with more capacity from hub to PC for downloads is acceptable. Category 3 or 5 unshielded twisted pair wiring is used for 100 Base-T4.

An alternative to 100 Base-T4 is 100 Base-TX. This uses two twisted pairs, with 100 Mbps in each direction. However, 100 Base-T4 has a signaling rate of only 125 MHz. It accomplishes this using Four Bit Five Bit (4B5B) encoding rather than Manchester encoding. Under 4B5B, every four bits is mapped into five bits in such a way that there are many transitions for digital receivers to lock onto, irrespective of the actual data stream. Since four bits are mapped into five bits, 4B5B is 80% efficient. Thus, 125 MHz times 0.8 yields 100 Mbps.

Finally, 100 Base-FX uses two strands of the lower performing multimode fiber. It has 100 Mbps in both directions and is for runs (say between buildings) of up to 2 km.

It should be noted that Fast Ethernet uses the signaling method for twisted pair (for 100 Base-TX) and fiber (100 Base-FX) borrowed from FDDI. The FDDI protocol was a 100 Mbps token ring protocol used as a backbone in the 1980s.

To maintain channel efficiency (utilization) at 100 Mbps, versus the original 10 Mbps, the maximum network size of Fast Ethernet is about ten times smaller than that of the original Ethernet.

2.4 Gigabit Ethernet

The ever growing amount of network traffic brought on by the growth of applications and more powerful computers motivated a revised, faster version of Ethernet. Approved in 1998, the next version of Ethernet operates at 1,000 Mbps or 1 Gbps and is known as Gigabit Ethernet, or 802.3z. As much as possible, the Ethernet committee sought to utilize existing Ethernet features.

Gigabit Ethernet wiring is either between two computers directly or, as is more common, in a star topology with a hub or switch in the center of the star. In this connection, it is appropriate to say something about the distinction between a hub and switch. A shared medium hub uses the established CSMA/CD protocol so collisions can occur. At most, one attached station can successfully transmit through the hub at a time, as one would expect with CSMA/CD. The half duplex Gigabit Ethernet mode uses shared medium hubs.

A "switch" on the other hand, does not use CSMA/CD. Rather, the use of buffers means multiple attached stations may send and receive distinct communications to/from the switch at the same time. The use of multiple simultaneous transmissions means that switch throughput is substantially greater than that of a single input line. Level 2 switches are usually implemented in software, level 3 switches implement routing functions in hardware [62]. Full duplex Gigabit Ethernet most often uses switches.

In terms of wiring, Gigabit Ethernet has two fiber optic options (1000 Base-SX and 1000 Base-LX), a copper option (1000 Base-CX) and a twisted pair option (1000-Base T).

The Gigabit Ethernet fiber option deserves some comment. It makes use of 8B10B encoding, which is similar in its operation to Fast Ethernet's 4B5B. Under 8B10B, eight bits (1 byte) are mapped into 10 bits. The extra redundancy this involves allows each 10 bits not to have an excessive number of bits of the same type in a row or too many bits of one type in each of 10 bits. Thus, there are sufficient transitions from 1 to 0 and 0 to 1 or the data stream even if the data has a long run of 1's and 0's.

Gigabit Ethernet using twisted pair uses five logic levels on each wire. Four of the logic levels convey data and the fifth is for control signaling. With four data logic levels, two bits are communicated at once or eight bits over all four wires at a time. Thus the signaling rate is 1 Gbps/8 or 125 MHz.

In terms of utilization under CSMA/CD operation, if the maximum segment size had been reduced by a factor of 10 as was done in going from the original Ethernet to Fast Ethernet, only very small gigabit networks could have been supported. To compensate for the ten times increase in data rate relative to Fast Ethernet, the minimum frame size for Gigabit Ethernet was increased (by a factor of eight) to 512 bytes from Fast Ethernet's 512 bits (see Robertazzi [52] for a discussion of the Ethernet equation that governs this).

Another technique that helps Gigabit Ethernet's efficiency is frame bursting. Under frame bursting, a series of frames are sent in a single burst.

Gigabit Ethernet's range is at least 500 m for most of the fiber options and about 200 m for twisted pair [62, 65].

2.5 10 Gigabit Ethernet

Considering the improvement in Ethernet data rate over the years, it is not too surprising that a 10 Gbps Ethernet was developed [61, 67]. Continuing the increases in data rate by a factor of ten that have characterized the Ethernet standards, 10 Gbps (or 10,000 Mbps) Ethernet is ten times faster than Gigabit Ethernet. Applications

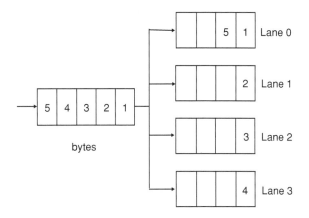

Fig. 2.3 Four parallel lanes for 10 gigabit Ethernet

include backbones, campus size networks and metropolitan and wide area networks. This latter application is aided by the fact that the 10 Gbps data rate is comparable with a basic SONET fiber optic transmission standard rate. In fact, 10 Gbps Ethernet will be a competitor to ATM high-speed packet switching technology. See the following chapters for more information on ATM and SONET.

There are eight implementations of 10 Gbps Ethernet. It can use four transceiver types (one four wavelength parallel system and three serial systems with a number of multimode and single mode fiber options). Like earlier versions of Ethernet, it uses CRC error coding. It operates in full duplex non-CSMA/CD mode. It can go more than 40 km via single mode fiber.

To lower the speed at which the Media Access Control (MAC) layer processes the data stream, the MAC operates in parallel on four 2.5 Gbps streams (lanes). As illustrated in Fig. 2.3, bytes in an arriving 10 Gbps serial transmission are placed in parallel in the four lanes.

There is a 12 byte Inter Packet Gap (IPG) which is the minimum gap between packets. Normally, it would not be easy to predict the ending byte lane of the previous packet, so it would be difficult to determine the starting lane of the next transmission. The solution is to have a starting byte in a packet always occupy lane 0. The IPG is found using a pad (add in extra 1–3 bytes), a shrink (subtract 1–3 bytes) or through combination averaging (average of 12 bytes achieved through a combination of pads and shrinks). Note that padding introduces extra overhead in some implementations.

In terms of the protocol stack, this can be visualized as in Fig. 2.4.

The PCS, PMA and PMD sub-layers use parallel lanes for processing. In terms of the sub-layers, they are:

Reconciliation: Command translator that maps terminology and commands in MAC into electrical format appropriate for physical layer.
PCS: Physical Coding Sublayer.
PMA: Physical Medium Attachment (at transmitter serialize code groups into bit stream, at receiver synchronization for data decoding).

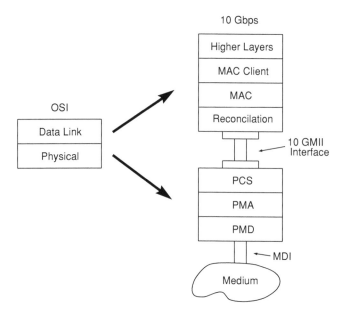

Fig. 2.4 Protocol stack for 10 Gbps Ethernet

PMD: Physical Medium Dependent (includes amplification, modulation, wave shaping).
MDI: Medium Dependent Interface (i.e. connector).

2.6 40/100 Gigabit Ethernet

Over the years Ethernet has been attractive to users because of its relatively low cost, robustness and its ability to provide an interoperable network service. Users have also liked the wide vendor availability of Ethernet-related products. However even with the release of gigabit and 10 Gbps Ethernet demand for bandwidth continued to grow. Network equipment shipments can grow at a 17% a year rate. Internet traffic grows at 75–125% a year. Computer performance doubles every 24 months. A 2008 projection was that within four years 40 Gbps would be needed.

One of the applications driving this growth is the increasing use of data centers (see the latter chapter on this topic). These facilities house server farms for hosting web services and cloud computing services. Projections indicate a need for 100 Gbps of data transfer capacity from switch to switch. Also 100 Gbps will have applications between buildings, within campuses and for metropolitan area networks (MAN) and wide area networks (WAN).

Table 2.4 40/100 Gbps
Ethernet

40 Gbps	100 Gbps
≥10 km Single-mode fiber	≥40 km Single-mode fiber
≥100 m Multi-mode fiber	≥10 km Single-mode fiber
≥10 m Copper cable	≥100 m Multi-mode fiber
≥1 m Backplane	≥10 m Copper cable

Fig. 2.5 40 and 100 Gbps
Ethernet protocol functions

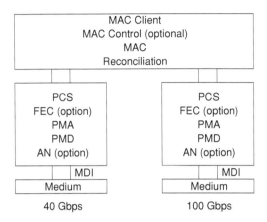

In July 2006 a committee was convened to explore the increasing data rate of Ethernet beyond 10 Gbps. In 2010 standards for 40 Gbps and 100 Gbps Ethernet were approved. This discussion is based on D'Ambrosia [11] and Nowell [40].

2.6.1 40/100 Gigabit Technology

In implementing 40 and 100 gigabit Ethernet some of the objectives are:

- Medium Access Control (MAC) data rates of 40 and 100 gigabit per second.
- Full duplex is only supported (i.e. two way communication).
- Maintain the existing minimum and maximum frame length.
- Use the current frame format and MAC layer.
- Optical transport network (OTN) support.

A variety of transmission media can carry 40 and 100 gigabit Ethernet as Table 2.4 illustrates.

Figure 2.5 illustrates the protocol stack for 40 and 100 gigabit Ethernet. In the figure one has the physical coding sublayer (PCS), the forward error correction sublayer (FEC), physical medium attachment sublayer (PMA), physical medium dependent sublayer (PMD) and the auto-negotiation sublayer (AN). Here also MDI is the medium-dependent interface or the connector.

In the physical coding sublayer 64B/66B coding is used, mapping 64 bits into 66 bits to provide enough transitions between 0 and 1 for digital receivers. As in

10 gigabit Ethernet, the concept of parallel lanes is used in 40 and 100 gigabit Ethernet. A 66 bit block is distributed in round robin fashion on the PCS lanes. Specifically for 40 gigabit Ethernet there are 4 PCS lanes that support 1, 2 or 4 channels or wavelengths. For 100 gigabit Ethernet there are 20 PCS lanes that support 1, 2, 4, 5, 10 or 20 channels or wavelengths. As an example, a 100 gigabit Ethernet may use 5 parallel wavelengths over a fiber, each carrying 20 Gbps.

Created PCS lanes can be multiplexed into any interface width that is supported. There is a unique lane marker for each PCS lane which is inserted periodically. Bandwidth for the lane marker comes from periodically deleting the inter-packet gap (IPG) in a lane. All bits in the same lane follow the same physical path no matter how multiplexing is done.

The receiver reassembles the PCS lanes by demultiplexing bits and also realigns the PCS lanes taking into account the skewness of the lanes. Advantages of this include the fact that all encoding, deskew and scrambling functions are implemented on a CMOS device located on the host and there is minimal bit processing except for using an optical module for multiplexing.

Finally, clocking takes place at 1/64th of the data rate (625 MHz for 40 gigabits and 1.5625 GHz for 100 gigabits). More information on 40 and 100 gigabit Ethernet can be found on http://www.ethernetalliance.org

2.7 Conclusion

For thirty years Ethernet has continually transformed itself by way of higher data rates to meet increasing demand for networking services. It will be interesting to see what the future holds.

Chapter 3
InfiniBand

3.1 Introduction

Data centers, facilities housing thousands of PCs, have become increasingly important for hosting web services, such as Google, cloud computing and hosting virtualized services. Data centers are also important for high performance computing
(HPC) facilities which turn the power of thousands of PCs loose on difficult but
important scientific and engineering computational problems. Data centers are an
impressive technological wonder, with their energy consumption being an important
issue affecting their location and cost of operation.

"InfiniBand," is a widely used interconnect used in high performance data centers
to connect the thousands of PCs to each other. The following discussion is based on
the more extensive treatment in Grun [17]. The product started to be offered in 1999.
Often InfiniBand transfers data directly from the memory of one computer to the
memory of another without going through the computer's operating systems. This
is referred to as "RDMA" or Remote Direct Memory Access. This is an extension
of Direct Memory Access (DMA) used in PCs. In DMA a DMA engine (controller)
allows memory access without involving the CPU processor. Off-loading this function to the DMA engine makes for a more efficient use of the CPU. The difference
between RDMA and DMA is that RDMA is done between remote (separate or
distant) machines whereas DMA is done on a single machine.

One of the simplifying features of InfiniBand is that it provides a "messaging
service" that applications can directly access. This compares to byte stream-oriented
TCP/IP over Ethernet, which is byte-oriented rather than message-oriented.

3.2 A First Look

The message service of InfiniBand is easy to use. It can allow communication from
an application to other applications or processes or to gain access to storage. In using
the messaging service the operating system is not needed. Instead an application

T. Robertazzi, *Basics of Computer Networking*, SpringerBriefs in Electrical
and Computer Engineering, DOI: 10.1007/978-1-4614-2104-7_3,
© The authors 2012

directly access the messaging service rather than one of the server's communication resources.

InfiniBand "creates" a channel application. Applications making use of the service can either be kernel[1] applications (such as file systems) or in user space. All of InfiniBand is geared toward supporting this top–down messaging service. Channels serve as pipes (i.e. connections) between disjoint virtual address spaces. These could also be disjoint physical address spaces (that is distinct servers separated by distance).

Queue Pairs

The end points of a channel are the send and receive queues. This is also known as a queue pair. When an application requires one or more connections more Queue Pairs (QP's) are generated. The Queue Pair maps directly into the virtual address spaces of each application. This idea is called "channel I/O."

Transfer Semantics

There are two ways in which data can be transferred in InfiniBand:

- SEND/RECEIVE: the application on the receiver side provides a data structure for received messages. The data structure is pre-posted on the receiver queue. Actually the sending side does not "see" the buffers or the data structure on the receiver side. The sending side just SENDS one or more messages and the receiver side RECEIVES them.
- RDMA READ/RDMA WRITE: the steps are as follows:

    (a) A buffer is registered in the receiver side application's virtual address space by the receiver side application.
    (b) Control of the buffer is passed to the sending side by the receiver.
    (c) The sending side uses the RDMA READ or RDMA WRITE operations to either read or write data in that buffer.

3.3 The InfiniBand Protocol

InfiniBand messages may be up to 2^{31} bytes. Messages are partitioned (segmented) into packets by the InfiniBand hardware. The packet size used is selected to make the best use of network bandwidth. InfiniBand switches and routers are used for transmitting packets through InfiniBand.

The generic OSI protocol stack layers that correspond to the InfiniBand messaging service is illustrated in Fig. 3.1. Here "SW" stands for software.

---

[1] A kernel is the most basic part of most operating systems. It connects applications to data processing in the computer's hardware. Kernel address space is used exclusively by the kernel, kernel extensions and most device drivers while user space is where the user applications reside and work.

Fig. 3.1 InfiniBand
equivalency to OSI protocol
stack (after Grun [17])

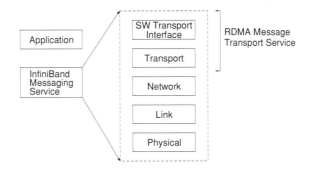

An InfiniBand switch is similar in theory to other types of common switches but is adapted to InfiniBand performance and cost goals. InfiniBand switches use cut through switching for better performance. Under cut through switching a node can start forwarding a packet before it is completely received by the node. InfiniBand link layer flow control is employed so during standard operation packets are not dropped. Ethernet has more loss than InfiniBand. It should be mentioned that, as opposed to switches, InfiniBand routers are not widely used.

Software for InfiniBand is made up of upper layer protocols (ULPs) and libraries. Mid-layer functions support the ULPs. There are hardware specific data drivers.

3.4 InfiniBand for HPC

The Message Passing Interface (MPI) is the leading standard and model for parallel system communication. Why use MPI? It offers a communication service to the distributed processes making up an HPC application. In fact MPI middleware[2] is used by InfiniBand. The MPI middleware in InfiniBand is allowed to communicate between machines in a cluster without the involvement of the CPUs in the cluster. The copy avoidance[3] architecture and stack bypass feature of InfiniBand provide extremely low application to application delay (latency), high bandwidth and low CPU loading.

Some other InfiniBand options, particularly for use with storage, include:

Socket Direct Protocol (SDP): this allows a socket application to use the Infini-Band network without changing the application.

SCSI RDMA Protocol: the Small Computer System Interface (SCSI) makes possible data transfer between computers and peripheral devices. It is commonly

---

[2] Middleware is software which connects application software to the operating system. Some types of middleware are eventually incorporated into newer versions of operating systems. An example of this migration is TCP/IP.

[3] Copy avoidance methods use less copying of data to memory by the operating system leading to higher data transfer rates. Zero copy methods make no copies.

pronounced "scuzzy." InfiniBand can enable a SCSI system to use RDMA semantics[4] to connect to storage.

IP over InfiniBand: this makes it possible for an application hosted by InfiniBand to communicate to the outside word using IP-based semantics.

NFS-RDMA: this is the Network File System over RDMA. The NFS is a widely used file system for use with TCP/IP networks. It allows a computer acting as a client to access files over a network in much the same way it access local files. It was originally developed by SUN Microsystems.

Lustre Support: Lustre is a large-scale (massive) file system for use in large cluster computers. Lustre can support tens of thousands of computers, petabytes of storage and hundreds of gigabytes/second of input/output throughput. It is used in both supercomputers and data centers (Wikipedia). It is available under GNU General Purpose License (GPL). The name "Lustre" comes from the words Linux and Cluster.

3.5 Conclusion

InfiniBand competes with Ethernet to provide high performance interconnects for HPC systems. Traditionally HPC was the primary market for InfiniBand. However it is finding a new role in cloud computing environments including those used for financial services where low latency and high bandwidth are important considerations.

---

[4] "Semantics" is the meaning of computer instructions—"syntax" is their format.

Chapter 4
Wireless Networks

4.1 Introduction

Wireless technology has unique capabilities to service mobile nodes and establish network infrastructure without wiring. Wireless technology has received an increasing amount of R&D attention in recent years. In this section, the popular 802.11 WiFi, 802.15 Bluetooth, 802.16 WIMAX and LTE standards are discussed.

4.2 802.11 WiFi

The IEEE 802.11 standards [16, 22, 25] have a history that goes back a number of years. The original standard was 802.11 (circa 1997). However, it was not that big a marketing success because of a relatively low data rate and relatively high cost. Future standardized products (such as 802.11b, 802.11a, and 802.11g) were more capable and much more successful. We will start by discussing the original 802.11 standard. All 802.11 versions are meant to be wireless local area networks with ranges of several hundred feet.

4.2.1 The Original 802.11 Standard

The original 802.11 standard can operate in two modes (see Fig. 4.1). In one mode, 802.11 capable stations connect to access points that are wired into a backbone. The other mode, ad hoc mode, allows one 802.11 capable station to connect directly to another without using an access point.

The 802.11 standard uses part of the ISM (Industrial, Scientific and Medical) band. The ISM band allows unlicensed use, unlike much other spectrum. It has been popular for garage door openers, cordless telephones, and other consumer electronics devices. The ISM band includes 902–928 MHz, 2.400–2.4835 GHz and 5.725–5.850 GHz. The original 802.11 standard used the 2.400–2.4835 GHz band.

T. Robertazzi, *Basics of Computer Networking*, SpringerBriefs in Electrical and Computer Engineering, DOI: 10.1007/978-1-4614-2104-7_4,
© The authors 2012

Fig. 4.1 Modes of operation
for 802.11 protocol

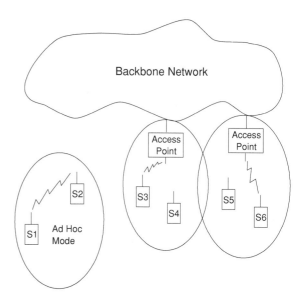

In fact, infrared wireless local area networks have also been built but are not used today on a large scale. Using pulse position modulation (PPM), they can support at least a 1–2 Mbps data rate.

The 802.11 standard can use either direct sequence or frequency hopping spread spectrum. Frequency hopping systems hop between 79 frequencies in the US and Europe and 23 frequencies in Japan. Direct sequence achieves data rates of 2 Mbps while frequency hopping can send data at 1 or 2 Mbps in the original 802.11 standard.

Because of the spatial expanse of wireless networks, the type of collision detection used in Ethernet would not work. Consider two stations, station 1 and station 2, that are not in range of each other. However, both are in range of station 3. Since the first two stations are not in range of each other, they could both transmit to station 3 simultaneously upon detecting an idle channel in their local geographic region. When both transmissions reach station 3, a collision results (i.e., overlapped garbled signals). This situation is called the hidden node problem [64, 65].

To avoid this problem, instead of using CSMA/CD, 802.11 uses CDMA/CA (Carrier Sense Multiple Access with Collision Avoidance). To see how this works, consider only station 1 and station 3. Station 1 issues a RTS (request to send) message to station 3 which includes the source and destination addresses, the data type and other information. Station 3, upon receiving the RTS and wishing to receive a communication from station 1, issues a CTS (clear to send) message signaling station 1 to transmit. In the context of the previous example, station 2 would hear the CTS and not transmit to station 3 while station 1 is transmitting. Note RTS's may still collide but this would be handled by rescheduled transmissions.

The 802.11 protocol also supports asynchronous and time critical traffic as well as power management to prolong battery life.

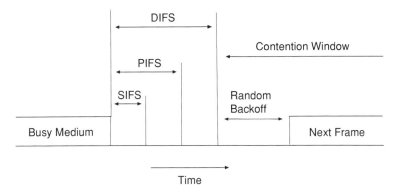

Fig. 4.2 Timing of 802.11 between two transmissions

Table 4.1 UNI bands

Name	Band
UNI-1	5.2 GHz
UNI-2	5.7 GHz
UNI-3	5.8 GHz

Figure 4.2 shows the timing of events in an 802.11 channel. Here after the medium becomes idle a series of delays called spaces are used to set up a priority system between acknowledgements, time critical traffic and asynchronous traffic. An interframe space is an IFS.

Now, *after* the SIFS (short interframe space) acknowledgements can be transmitted. After the PIFS (point coordination interframe space) time critical traffic can be transmitted. Finally, after the DIFS (distributed coordination interface space) asynchronous or data traffic can be sent. Thus, acknowledgements have the highest priority, time critical traffic has the next highest priority, and asynchronous traffic has the lowest priority.

The original 802.11 standard had an optional encryption protocol called WEP (wired equivalent privacy). A European competitor to 802.11 is HIPERLAN (and HIPERLAN2). Finally, note that wireless protocols are more complex than wired protocols, for local area network and other environments.

4.2.2 Other 802.11 Versions

Since the original 802.11, a number of improved versions were developed and have become available. The original 802.11 version itself did not sell well as the price and performance was not that appealing. The four general variations are 802.11b, 802.11a, 802.11 g and 802.11n. Each is now briefly discussed. See Kapp [22], Kuran [23], Poole [47], [48], Vaughan-Nichols [67], [68] for discussions. There

Table 4.2 Other 802.11 protocols

Name	Description
802.11e	With quality of service
802.11h	Standard for power use and radiated power
802.11i	Uses WEP2 or AES for improved encryption
802.11p	Vehicular environments
802.11s	Mesh technology support
802.11u	Inter-networking support for external networks
802.11x	Light weight version of EAP (Extended Authentication Protocol)

are 800 million new 802.11 devices sold every year [69]. More US households use WiFi for a home LAN rather than Ethernet.

802.11b: This 1999 version first made WiFi popular. It operates at a maximum of 11 Mbps at a range of 100–150 feet and 2 Mbps at 250–300 feet. Data rate decreases with distance to 1 Mbps and then goes to 0. If Wired Equivalent Privacy is used with encryption, the actual useful data rate drops by 50%.

The 802.11b signal is in the 2.4 GHz band. It can operate either using direct sequence spread spectrum, frequency hopping or infrared. Direct sequence is very popular and infrared is mostly not in use.

802.11a In spite of the name, 802.11a was developed after 802.11b. It operates at 54 Mbps in the Unlicensed Infrastructure Band (UNI, Table 4.1):

There is some disagreement in the technical literature as to whether 802.11b or 802.11a has the larger range.

802.11g: Sometimes 802.11g is known as 802.11b extended. Initial versions were at 22 Mbps, later versions were at 54 Mbps. Using 802.11g and methods to increase throughput can offer data rates of 100–125 Mbps [23].

802.11n: A committee was formed to create a higher rate version of 802.11 [47, 48] in 2004. Two years later there was early industry agreement on the features of 802.11n. The most significant feature of 802.11n is the use of MIMO (multiple input multiple output) antenna technology. Normally when a radio signal is transmitted from a transmitter to a receiver, the receiver may receive multiple time shifted versions of the original signal. This is because the signal may bounce off different objects and be reflected to the receiver along paths of different lengths. This is called multi-path interference. While normally a nuisance, MIMO uses multiple antennnas to achieve the transmissions of multiple spatial channels in parallel. This boosts throughput. The throughput of 802.11n can be as high as 600 Mbbps but this is with 40 MHz channels (20 MHz and 40 MHz channels can be used) and with four spatially parallel streams. The maximum range of 802.11n is 50 m. The modulation scheme used is OFDM (orthogonal frequency division multiplexing).

Note access points can include options for more than one 802.11 version.

A number of specialized 802.11 standards have also been in development. Some are given in Table 4.2.

One of these that it would be good to discuss is:

802.11s: Traditionally devices such as laptops are wirelessly connected to access points. These access points are hard wired into the external Internet. The standard 802.11s provides support to allow wireless units such as access points to form wireless mesh networks [15,19, 69]. This is support. The actual communication is still handled by 802.11 a,b,g or n. A working group was set up by IEEE to create this standard in 2004. There have been thousands of comments and several delays and a stable version was created around 2010 [15].

In 802.11s the basic unit is a "mesh station" which is networked with other mesh stations statically or in an ad hoc fashion. A collection of mesh stations and the networked links between them is a mesh basic service set (MBSS). A mesh station can be a gateway or bridge to an external network in which case it can be referred to as a mesh portal.

Techniques are used in 802.11s that hide the multi-hop nature of the mesh from the upper layers of the protocol stack. It seeks to be functionally equivalent to a broadcast Ethernet.

There is a mandatory path (route) selection algorithm and a mandatory link metric for inter-operability reasons but other algorithms/metrics can be utilized. Unicast, multicast and broadcast transmission of frames are supported in 802.11s. Mesh stations can detect each other using passive scanning (noticing beacon frames) or active scanning (sending probe frames). Stations may user single transceivers (on one frequency) or multi-transceivers (on different frequencies). Finally most 802.11s meshes will use the 802.11u standard for authentication of users.

A word is in order on 802.11 security. A user requires some sophistication to prevent snooping by others. For instance, security features on shipped products are often disabled by default. Williams reports that many corporate users are not using or misusing WEP. These are media articles on people driving by in vans tapping into private networks. The 2001 article by Williams describes a series of security weaknesses in 802.11b. Some in the wireless LAN industry feel if one uses accepted security practices along with 802.11 features, security is acceptable.

4.3 802.15 Bluetooth

The original goal of Bluetooth technology, standardized as IEEE 802.15, is to provide an inexpensive, low power chip that can fit into any electronic device and use ad hoc radio networking to make the device part of a network. For instance, if your PC, printer, monitor and speakers were Bluetooth enabled, most of the rat's nest of wiring under a desk top would be eliminated. Bluetooth chips could also be placed in PDAs, headsets, etc.

Work on Bluetooth started in 1997. Five initial corporate supporters (Erickson, Nokia, IBM, Toshiba and Intel) grew to a more than a thousand adopters by 2000. The name Bluetooth comes from the Viking King of Denmark, Harald Blatand, who unified Norway and Denmark in the tenth century [18].

4.3.1 Technically Speaking

Bluetooth had a number of design goals. As related in Haartsen [18], among these are:

• System should function globally.
• Ad hoc networking.
• Support for data and voice.
• Inexpensive, minature and low power radio transceiver.

The original Bluetooth operates in the 2.4 GHz ISM band (see the previous section's discussion of the ISM band). It uses frequency hopping spread spectrum (79 hopping channels, 1600 hops/s). Time is divided into 625 µs slots with one packet fitting in one slot. The data rate is 1 Mbps. The range is 10 m, making Bluetooth a personal area network (PAN).

Two types of connections are possible with Bluetooth. First, SCO links (Synchronous Connection Oriented) are symmetrical, point to point, circuit switched voice connections. Second ACL links (Asynchronous Connectionless) are asymmetrical or symmetrical, point to multipoint, packet switched connections for data.

A number of features of Bluetooth are designed to make possible good interference immunity. One is the use of high rate frequency hopping with short packets. There is an option to use error correction and a fast acting automatic repeat request scheme using error detection. Finally, voice encoding that is not that susceptible to bit errors is used.

4.3.2 Ad Hoc Networking

Two or more Bluetooth nodes form a "piconet" in sharing a frequency hopping channel. One node will become a "master" node to supervise networking. The other nodes are "slaves". Not only may roles change but roles are lost when a connection is finished.

All SCO and ACL traffic is scheduled by the master node [18]. The master node allocates capacity for SCO links by reserving slots. Polling is used by ACL links. That is, the master node prompts each slave node in turn to see if it has data to transmit (see Schwartz [56] for a detailed discussion of polling). Slave node clocks are synchronized to the master node's clock.

There is a maximum of eight active nodes on a single piconet (others may be parked in an inactive state). As the number of nodes increases, throughput (i.e. useful information flow) decreases. To mitigate this problem, several piconets with independent but overlapping coverage can form a "scatternet". Each piconet that is part of a scatternet uses a separate pseudo-random frequency hopping sequence. The scatternet approach results in a very small decrease in throughput. Note that a

Table 4.3 Bluetooth versions [69]

Version	Data rate	Max application thruput
Version 1.2	1 Mbps	0.7 Mbps
Version 2.0+	3 Mbps	2.1 Mbps
Version 3.0+	maybe 24 Mbps	2.1 Mbps
Version 4.0	maybe 24 Mbps	2.1 Mbps

node can be on several piconets in a scatternet. Moreover, it may be a master node on some piconets and a slave node on other piconets.

4.3.3 Versions of Bluetooth

There are four versions of Bluetooth as illustrated in Table 4.3.

The indicated 24 Mbps data rate in the table is not transmitted over Bluetooth itself. Rather it is transmitted over a parallel 802.11 link whose operation is negotiated using Bluetooth. Version 4.0 includes protocols for (a) classic Bluetooth, (b) Bluetooth high-speed and (c) Bluetooth low energy.

There was some effort that was not realized to make an 802.15 variant ultra wide band (UWB) standard for Bluetooth [30]. Ultra wide band technology is a short range radio technology that spreads the communication spectrum over an unusually wide band of spectrum. It can have low energy requirements.

4.3.4 802.15.4 Zigbee

In actuality, the original Bluetooth faced some problems in gaining acceptance. The rapid growth of 802.11 technology and its pricing had not given Bluetooth a price advantage on certain applications [72]. Also, Bluetooth is more complex than its original design goal as an attempt was made to have it serve more applications and supply quality of service. There is also some question on the scalability of scatternets (Liu).

Two planned successors to the original 802.15 Bluetooth standard were 802.15.3a for high rate ultra wideband (UWB) wireless personal area networks (WPAN) and 802.15.4 for low rate, low power WPAN's. In this section, the low data rate extension of Bluetooth is discussed.

A great many applications could benefit from a low data rate Bluetooth standard [72]. Among these are home, factory and warehouse automation. These are applications for monitoring involving safety, the health field and the environment. The use of low data rate Bluetooth for precision asset location and situation awareness could take place for emergency services and inventory tracking. Finally, there are potential entertainment applications including interactive toys and games.

Table 4.4 Zigbee channels

No. of channels	Data rate (Kbps)	Band
16	250	2.4 GHz
10	40	915 MHz
1	20	868 MHz

Zigbee can operate either in the 2.4 GHz ISM band (available worldwide), the ISM 868 MHz band (Europe) and the ISM 915 MHz band (North America). Twenty-seven channels are defined for 802.15.4 as indicated in Table 4.4.

Zigbee, like Bluetooth, has a range of 10 m. Communication can take place from a device to a coordinator, a coordinator to a device or between stations of the same type (i.e. peer to peer) in a multi-hop mode of operation. An 802.15.4 network can have up to 64,000 devices in terms of address space. Zigbee topology includes a one hop star or the use of multi-hopping for connectivity beyond 10 m.

In beacon enabled mode, the coordinator periodically broadcasts "beacons" to synchronize the devices it is connected to and for other functions. In non-beacon enabled mode, beacons are not broadcast periodically by the coordinator. Rather, if a device requests beacons, the coordinator will transmit a beacon directly to the device [72]. A loss of beacons can be used to detect link or node failures.

It is critical for certain applications to minimize Zigbee coordinator and device energy usage. Some of these applications will be battery powered where batteries will not be (practically or economically) replaceable.

The majority of power savings functions in 802.15.4 involve beacon-enabled mode. In direct data transmissions between coordinators and devices, the transceivers are only on 1/64 of the duration of a packetized superframe (i.e. collection of slots). A small CSMA/CD backoff duration, and brief warm-up times for transceivers are also used to minimize power usage in 802.15.4.

Three security levels are available in 802.15.4. The lowest level is None Security mode which is suitable if the upper layers provide security or security is not important. An access control list is used in the second level of security to allow only authorized devices to access data. The Advanced Encryption Standard (AES) is used in the highest, third security level.

Zigbee is built on an IEEE 802.15.4 foundation. While it is widely used in applications involving metering it is not suitable for industrial applications with real time requirements. Two industry standards also built on 802.15.4 are Wireles HART and ISA100 [71]. They emphasize real time and co-existence functionality.

4.3.5 802.15.6 Wireless Body Area Networks

Wireless body area networks (WBAN) are networks that can collect and transmit data from different parts of the human body, possibly to remote sites [24, 26]. Applications include medicine, finding lost items, data file transfer, gaming (sensors can collect data on body part movement) and social networking (for instance exchanging electronic business cards by simply shaking hands).

Medicine may be the biggest application of wireless body area networks. An obvious application is to collect vital patient data and forward it to a remote site for processing and diagnosis. This could be used for myocardial infarctions, and also for diseases involving cancer, asthma, the neural system and the gastrointestinal tract [24].

Work on the IEEE 802.15.6 standard is an attempt to standardize networking for wireless body area networks.

As of about 2010, 802.15.6 provides three physical (PHY) layer specifications [24]. The use of each depends on the type of application need:

(1) Narrowband PHY: This physical layer handles radio transceiver activation and deactivation, data transmission and receiving and clear channel assessment within the current channel.
(2) Ultra Wideband PHY: This has a low and high band. There are eleven potential channels of 499.2 MHz bandwidth each. The low band consists of channels 1 through 3 and the high band consists of channels 4 through 11. Channel 2 has a center frequency of 3.9936 GHz and channel 7 has a center frequency of 7.9872 GHz. Channel 2 and channel 7 are "mandatory" channels, other channels do not have to be used. Typical data rates are 0.5–10 Mbps.
    Ultra wideband transceivers are relatively simple.
(3) Human Body Communication PHY: This has two frequency bands of 4 MHz bandwidth and center frequencies of 16 MHz and 27 MHz. Both bands can be used in the US, Korea and Japan. The 27 MHz band can be used in Europe.

A super-frame[1] structure is used by 802.15.6. Beacon periods (of the same length as a super-frame) bound each super-frame. Boundaries of the beacon period are selected by the hub station which is thereby allocating time slots.

Like Zigbee, with 802.16.6 there are security options for (a) unsecured communication (b) authentication only and (3) authentication and encryption.

4.3.6 Bluetooth Security

Bluetooth has been is use for about 10 years. Hackers and researchers have discovered several security weaknesses during this time in Bluetooth enabled devices [12]. In fact most Bluetooth attacks are not detected and are more localized than Internet attacks so that they do not get the same amount of attention from the public. It should also be noted that mobile and embedded Bluetooth devices that receive attacks have few or no security features.

Dunning [12] holds that one should use Bluetooth if proper security is in place.

Most of the susceptibility of Bluetooth devices to attacks comes from lax default security settings, a lack of understanding on the part of Bluetooth device owners of

---

[1] A super-frame is a sequential concatenation of frames that repeats periodically.

security practice and deficient software development. One is safe from most Bluetooth attacks if security settings are correctly configured.

Bluetooth has a few weaknesses that are inherent [12]:

- Wireless data can be (locally) intercepted.
- No third party can verify addresses, names and classes as on the Internet.
- Many devices cannot be patched so any weaknesses remain as long as they are in use.

While Bluetooth threats will probably increase, the key to making them less effective is a better understanding of their potential.

4.4 802.16 Wireless MAN

4.4.1 Introduction

While wireless connectivity is very convenient, 802.11 and 802.15 have somewhat limited ranges (hundreds of feet and 10 m, respectively). A third wireless standard is IEEE 802.16 [1, 4, 13, 44, 46, 64, 65]. It defines how a base station may provide connectivity to computers in buildings up to several kilometers distant, or more recently, to mobile users. A home or business owner may install an antenna that allows him or her broadband Internet connectivity while bypassing telephone or cable company wired services. A mobile user's laptop or cell phone may also achieve connectivity using 802.16.

4.4.2 The Original 802.16 Standard

Standard efforts on 802.16 began in 1999 and the original standard was published in April 2001. The original standard was only for fixed location users.

The original 802.16 operates in the 10–66 GHz band to make possible a large bandwidth and thus data rate. Precipitation can be a problem on this band so forward error correction (Reed Solomon) coding is used. There is also an option for the use of CRC coding. Radio at 10–66 GHz is directional so base stations can have several antennas, each covering a different geographic sector.

The original 802.16 standard calls for the use of either time division duplexing (TDD: base station and user antenna share a single channel divided into time slots but do not transmit concurrently) or frequency division duplexing (FDD: separate channels, sometimes transmit concurrently). Note that in TDD there is a variable upstream/downstream capacity allocation via time slot assignment.

These types of modulation are used in the original 802.16 depending on distance (Table 4.5):

Table 4.5 802.16 Modulation Formats

Distance	Modulation	Bits per Symbol	No. of Waveforms
Small Distance	QAM-64	6 bits/symbol	64
Medium Distance	QAM-16	4 bits/symbol	16
Large Distance	QPSK	2 bits/symbol	4

Fig. 4.3 Protocol stack for the 802.16 standard

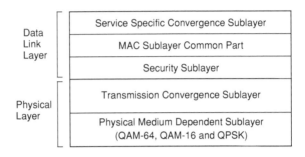

In general, a digital modulation scheme encodes a symbol as one of a number of possible amplitudes and phase shifts. For instance, QAM-64 has 64 combinations of amplitude and phase shifts so the equivalent of 6 bits ($2^6 = 64$) can be transmitted per symbol. As distance increases, it is harder to distinguish between the 64 combinations due to noise and channel effects so that fewer, more distinctive combinations are used (16 for QAM-16 and then 4 for QPSK) with corresponding lower bit/symbol rate. Note that if X MHz of spectrum is available, the data rates are 6X Mbps for QAM-64, 4X Mbps for QAM-16 and 2X Mbps for QPSK.

In terms of security, mutual authorization using RSA public key cryptography and X.509 certificates is specified in the original standard.

A variety of downstream traffic can be supported in connection-oriented mode. The four supported classes are:

• Constant bit rate.
• Real time variable bit rate.
• Non-real time variable bit rate.
• Best effort.

The physical and data link protocol layers for 802.16 appear in Fig. 4.3.

The service specific convergence sublayer interfaces to the network layer and is similar to the LLC (Logical Link Control) layer in older 802 standards. This includes interfacing for ATM (ATM convergence sublayer) and IPv4, IPv6, Ethernet and VLAN (packet convergence sublayer). The channel is supervised by the MAC sublayer common part. Security (encryption, key management) is the responsibility of the security sublayer.

Table 4.6 Some 802.16 Standards

Standard	Notes
802.16-2001	First standard
802.16-2004	OFDM and OFD-MA physical layer support
802.16e-2005	Mobile station connectivity
802.16f-2005	Management information base (MIB) for 802.16-2004
802.16g-2007	Management plane procedures and services
802.16j-2009	Multi-hop relay
802.16-2009	Roll up of some earlier standards for fixed and mobile users
802.16m	For 4G requirements
802.16n	Higher reliability nets

4.4.3 More Recent 802.16 Standards

Since the original 2001 802.16 standard, there have been a number of improved 802.16 standards developed. For instance, the 802.16e standard of 2005 allowed connectivity to mobile users.

Flexibility is a key feature of the 802.16 standardization effort. Several parts in fact are optional. To be more specific, the standards deal largely with the physical (PHY) and media access control (MAC) layers only and do not specify upper layer signaling and the network architecture. This was a major reason that the WIMAX forum was established in 2001 (http://www.wimaxforum.org). The WIMAX forum seeks to enable conformity and interoperability. Table 4.6 illustrates some of the 802.16 standards.

Two of these standards that would be interesting are discuss below.

4.4.4 802.16j

This is a standard that allows "relays" to be inserted within a base station's coverage region to use multi-hop technology to assist communication between a mobile station (MS) and a base station (BS). Such networks are referred to as multi-hop cellular networks. A task force to develop this standard was created in 2006.

There are a number of very interesting uses for relays in a cellular context. Among them are [46]:

- *Fixed Infrastructure:* Here relays are placed by a service provider in a fixed area. These are probably placed above rooftop height. Also possible are commercial relays owned by subscribers that may enter or leave the network.
- *In-Building Coverage:* The interior of buildings sometimes have "coverage holes". A service provider can locate relays near the exterior or just inside buildings to mitigate this problem. Relays can also be used in the vicinity of subways or tunnels.
- *Transient Coverage:* Relays can be used to provide additional connectivity for large and dense groups of people meeting in temporary situations. Relays in this

Fig. 4.4 The IEEE 802.16m
data and control plane

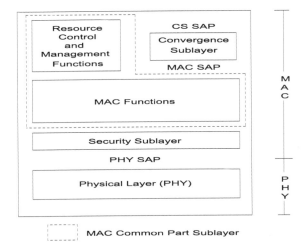

☐ MAC Common Part Sublayer

context can allow some traffic to be sent to base stations in adjacent cells. Relays can also be used to maintain coverage in case of base station damage.

- *Vehicular Coverage:* This use includes buses and trains. Design of such relay systems is a challenging problem as a mobile subscriber may hear signals both from the relay and a base station. In general a relay transmits on both the uplink and the downlink.

4.4.5 802.16m

The 802.16m standard will be considered for global 4G cellular network standardization (i.e. IMT-Advanced). The goal of 802.16m is to make use of 802.16-2009 and 802.16j-2009 to create a capable air interface and to support legacy equipment. The 802.16m standard introduces [4]:

- *Femto Cell Support:* Femto cells are small cells for the small office home office (SOHO) environment. The femto cell base station is connected to the Internet through a standard wired connection.
- *Self-Organizing Networks:* The purpose is to provide self organizing network functionality to allow automatic configuration of base station parameters. The amount of functionality should increase with more standard work.
- *Self-Optimization Procedures:* These allow a tuning of base station performance parameters based on processing self-organizing network measurement data.

The physical layer of 802.16m uses OFDMA (Orthogonal Frequency Division Multiple Access) and is to work in licensed bands under 6 GHz. Data rates for 802.16m are 100 Mbps for mobile users and 1 Gbps for fixed users.

A diagram of the 802.16m data and control plane appears in Fig. 4.4.

4.5 LTE: Long Term Evolution

4.5.1 Introduction

Long Term Evolution or "LTE" is a process to create a new air interface for cell phones by the 3rd Generation Partnership Project (3GPP). However LTE can also be considered to be a system consisting of architecture and protocols and performance goals.

When cell phones first came into use in the early 1980s these "1st generation" phones were analog based. Second generation phones were digital. Third and fourth generations have increasing data rates allowing new services such as web surfing. Some people call LTE a 3.9 technology in that it is almost but not quite 4th generation technology. But the more recent "LTE Advanced" is a 4th generation system.

The initial proposal for LTE came from NTT DoCoMo of Japan. Began first as a study item, the technical requirements of LTE were agreed to in 2005. The first working LTE service for the public was in Stockholm and Oslo in 2009. A number of carriers publicized plans, starting in 2009, to transform their networks into LTE networks. These efforts began in 2010. Finally, LTE Advanced was submitted as a candidate system that is fourth generation to ITU-T (International Telecommunications Union - Standardization Sector). The plan was for this to be finalized by 3GPP in 2011 [69].

4.5.2 LTE

The goals of LTE include [47, 48, 50] reduced latency, higher data rates for customers, better system capacity and coverage and a lower cost of operation.

Some important LTE features include [2, 50]:

- A flat IP oriented network architecture using distributed servers.
- In LTE, base stations have transport connectivity to the core network without intervening RAN (radio access network) nodes (e.g. radio network controllers).
- Elegant and efficient radio protocols. State information on channels is accessible to radio protocols peers for efficient operation.
- A physical layer design based on processing in the frequency domain (Fast Fourier Transforms (FFTs) are used). This aids efficient operation and helps to support high data speeds. Bins with a 20 MHz width are used with a 2048 point FFT computation.
- The use of multiple antenna transmission.
- Clever resource management of the radio spectrum to enhance scalability of bandwidth and multi-user diversity. For instance, scheduling can be done in the time-frequency domains.
- Power saving mode is an inherent feature of customer equipment.

In terms of some of the system parameters of the LTE system one has [47, 48, 69]:

- A peak download rate of 326.4 Mbps for 4 × 4 antennas and a peak download rate of 172.8 for 2 × 2 antennas, both utilizing 20 MHz of spectrum. A peak upload rate of 86.4 Mbps for each 20 MHz of spectrum (with one antenna).
- At least 200 active users in each 5 MHz cell.
- Latency below 5 msec for short IP packets.
- In rural regions cell sizes of 5 km, 30 km and 100 km. For urban regions cell sizes cam be as small as 1 km or less.
- There are five terminal classes. These include one class that emphasizes voice and a high end terminal class supporting maximum data rates.
- LTE has excellent support for mobility. High data rates are possible at speeds up to 300–500 km/h. This is a function of the frequency band employed.

4.5.3 LTE Advanced

Following LTE is the fourth generation radio technology standard called LTE Advanced [2, 69]. Another name for LTE Advanced is LTE Release 10. Fundamentally LTE Advanced is intended to allow for higher data rates and transmission speeds for cellular telephone networks. In fact LTE Advanced is backwards compatible with LTE. It uses the same frequency bands as LTE. However LTE is not compatible with other third generation systems.

Since LTE uses a 2048 point FFT for 20 MHz bins, it is reasonable to expect that LTE Advanced with 100 MHz bins uses 10,240 point FFTs [2]. The high mobility data rate for LTE Advanced is 100 Mbps and the low mobility data rate is 1 Gbps. As one might expect, LTE Advanced needs smaller latencies than LTE.

One advantage of LTE advanced is its capability to make use of networks consisting of macro cells and low power nodes (i.e. picocells, femtocells[2] and new relay nodes). Orthogonal frequency division multiplexing (OFDM) is used in LTE Advanced as well as multiple antenna techniques.

---

[2] A picocell is a cellular telephone base station covering a small indoors area such as a shopping mall, bus station or plane. A femtocell is a cellular telephone base station covering an even smaller area such a house or small building. A picocell has a range of about 200 m and a femtocell has a range of tens of meters [69].

Chapter 5
Asynchronous Transfer Mode (ATM)

5.1 Asynchronous Transfer Mode (ATM)

The technology of asynchronous transfer mode (ATM) was developed in the telephone industry in the 1980s at the major telephone research laboratories. The goal
was to develop a networking technology that could carry an integrated mix of traffic
(voice, video and data). At one point, there were two competitors to be the technology of choice, synchronous transfer mode (STM) and ATM. Asynchronous transfer
mode was the eventual winner.

In fact, ATM is a packet switched-based technology using virtual circuits.
That is, all of the packets in a flow between two nodes follow the same path. Fixed
length packets are used (53 bytes: 5 bytes of header and 48 bytes of data). Special
architectures for high speed, high throughput, ATM packet switching have been
developed and are discussed below.

While a number of small companies attempted to market ATM local area networks,
ATM found a place in the backbones of the Internet. It is not clear that it will have
staying power. Some effort is underway to eliminate what some see as an unnecessary
layer of complexity in protocol stacks.

Before discussing some aspects of ATM, let us first contrast it with STM.

5.1.1 Limitations of STM

Conceptually, STM [31] is closer to traditional telephone circuit switching than ATM.
The basic idea is to build a high speed channel out of a small set of basic channel
building blocks. For instance, a need for a 200 Kbps channel is met by aggregating
four standard 64 Kbps digital voice channels (called B channels). Or a need for
4 Mbps is met by aggregating three 1.536 Mbps H1 channels.

T. Robertazzi, *Basics of Computer Networking*, SpringerBriefs in Electrical
and Computer Engineering, DOI: 10.1007/978-1-4614-2104-7_5,
© The authors 2012

Table 5.1 Burstiness of traffic

Class	Peak rate (kbps)	Peak/mean bit rate
Voice	16–64	2:1
Text	1–64	2:1
Image/Data	64–2,000	10:1
Video	$\leq 140,000$	5:1

What were some of the problems with STM that led to its abandonment?

- The hierarchy is rigid. Note that in the two examples above, because there are only a small number of basic channel data speeds, in most cases some capacity is wasted once channels are aggregated.
- As in digital telephony, switching is done by time slot mapping. But STM leads to a complex time slot mapping problem.
- It was also felt that multi-rate switching was difficult to build out of 64 kbps building blocks.
- Separate switches may be needed for each type of traffic to be carried. This is a real show stopper. Traffic prediction, especially by class, is inexact and one could install many switches of some classes that could be under or over-utilized. In a national network, the economic loss could be very large.
- Circuit switching is not efficient for bursty (i.e. intermittent) data traffic (see Table 5.1) as capacity is allocated even if at that instant no data is transmitted. This is a waste of resources. It may not be practical to set up a channel only when bursty data is present due to the overhead of call set up time.

Packet switching and ATM, on the other hand, are particularly suited for bursty traffic. There is what is called statistical multiplexing and inherent rate adaptation. That is, links and other resources are efficiently shared by multiple source/destination pairs. If some sources are idle, other busy sources can easily make use of link capacity.

Moreover, there is no need for different types of switches for each service class. From the viewpoint of the 1980s, one could justify the investment in a single type of switch on the majority voice traffic (that has since changed) and experimentally run new services on top of that at relatively low marginal cost. In fact, in a sense the design problem for a single type of switch is simply to design the highest throughput switch for the amount of money available. In fact, things are really a bit more complex (there is quality of service, for instance) but the design problem is still simpler compared to the STM alternative.

5.1.2 ATM Features

Once again the ATM packet (called a "cell" in ATM language) is 53 bytes long, 5 bytes of which are header. There are actually two types of headers, depending on whether the packet is traversing a link at the boundary of the ATM network (UNI: user to network interface) or a link internal to the ATM network (NNI: network to network interface). See Fig. 5.1a for an illustration [41]. In ATM language, a generic

Fig. 5.1 An (**a**) ATM
network and (**b**) a
transmission path broken
into virtual paths and virtual
channels

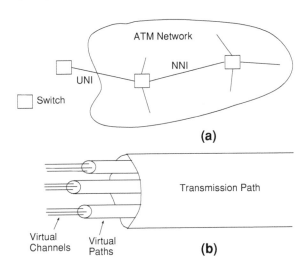

virtual circuit is called a "virtual channel". Virtual channels are bundled into "virtual paths". A link carries multiple virtual paths, each consisting of many virtual channels (see Fig. 5.1b).

Let us examine the UNI header first (Fig. 5.2). The first four bits are for generic flow control (GFC), a field that was incorporated in the header but in reality is not used. There are eight bits for virtual paths (thus, $2^8 = 256$ virtual paths/link) and 16 bits for virtual channels (thus, $2^{16} \cong 64,000$ virtual channels/virtual path). The 3 bit payload type (PT) field indicates eight possible payload types. The 1 bit cell loss (CL) priority field indicates whether or not the cell can be dropped under congestion. Finally, the header error check field uses a code to protect the header only, with single bit error correction and detection for 90% of multiple bit errors. Note that if error protection is needed for the data, this has to be taken care of at a different protocol layer.

The NNI header fields are similar to the UNI fields except that there is no generic flow control field in the NNI header and 12, rather than 8, bits are reserved for virtual paths. Thus, the NNI has $2^{12} = 4,096$ virtual paths compared to the $2^8 = 256$ virtual paths for the UNI. This is because of the NNI links are more like internal trunks which carry more traffic than the access links, as with the UNI. Note that since the number of bits in the virtual channel field is the same for both the UNI and the NNI, a single virtual path contains the same number of virtual channels (2^{16}) whether it involves the UNI or NNI.

The 53 byte packet size for ATM was chosen as a compromise among interested parties and partly serves to minimize queuing delay through the use of a relatively short packet.

A word is in order on the payload type field. The eight (2^3) possibilities are given in Table 5.2 [65].

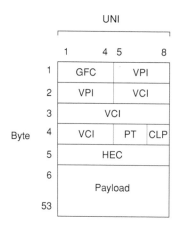

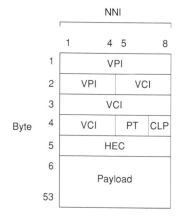

UNI: User to Network Interface

NNI: Network to Network Interface

GFC: Generic Flow Control

VPI: Virtual Path Identifier

VCI: Virtual Channel Identifier

PT: Payload Type (3 bits)

CLP: Cell Loss Priority (1 bit)

HEC: Header Error Check

Fig. 5.2 UNI and NNI cell headers

Table 5.2 Payload Types

Type	Explanation
000	User data cell, no congestion, packet type 0
001	User data cell, no congestion, packet type 1
010	User data cell, congestion present, packet type 0
011	User data cell, congestion present, packet type 1
100	Management info for adjacent ATM switches
101	Management info for source/destination ATM switches
110	Resource management cell
111	Reserved for future use

It is a bit redundant, but cell type (whether or not a packet can be dropped under congestion) can be indicated either through the payload type or cell loss priority field.

Special resource management (RM) cells are inserted periodically on virtual channel streams for congestion control and then return to the source. An RM cell that does not return to the source in a reasonable time indicates congestion. An explicit transmission rate in the RM cell can also be lowered by congested switches to throttle back the source. Overloaded switches can also create RM cells. This rate based congestion control is used in ATM but a good discussion of other discarded possibilities for congestion control appears in the 3rd edition of Tannenbaum [64].

In order to carry a mix of traffic, ATM supports four classes of traffic. They are:

- Constant bit rate (CBR).
- Variable bit rate (VBR), which consists of real time (RT-VBR) and non-real time (NRT-VBR) option.
- Available bit rate (ABR).
- Unspecified bit rate (UBR).

Available bit rate class may guarantee a minimum data rate but exceeds that sometimes. Unspecified bit rate class gives no guarantees and can be used for file transfers or email.

ATM technology can provide quality of service (QoS) guarantees. Two communicating nodes agree (through a "contract") on QoS parameters specifications. There are a large number of QoS parameters, such as minimum and peak cell rate, cell transfer delay, cell error rate as well as others.

5.1.3 ATM Switching

A "switch" is a computerized device that interconnects users or computers. In other words, a switch provides connectivity. Switching can in general be based on either circuit switching or packet switching. There are three general architectures for building ATM packet switches: shared medium, shared memory and space division switching [51]. Each is now discussed in turn.

Shared Medium Architecture

This architecture uses a computer bus to provide the interconnection capability. An illustration appears in Fig. 5.3. A computer bus is a fairly large number of parallel wires. While each operates at a moderate speed, the aggregate rate is impressive. For instance, a 64 bit bus with each wire at 80 Mbps has an aggregate data rate in the gigabits per second (i.e. $64 \times 80 \times 10^6$ bps).

A number of buffers provide input and output paths to the bus. The buffers exiting the bus are shown larger than the input buffers as it is usual for the aggregate speed of the bus to be significantly larger than the input port rate. A bus controller coordinates access to the bus.

A little thought will show that for a switch with N inputs, the bus should operate at least N times faster than the input packet arrival rate. Thus, the switch is said to have an N times speed up.

Shared Memory Architecture

A shared memory switch architecture is shown in Fig. 5.4. The inputs are multiplexed into a single stream accessing a dual port memory. The memory is organized into packet buffers and address chain pointers.

Fig. 5.3 Shared medium
switch architecture

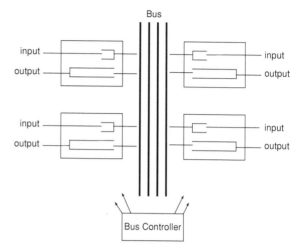

Fig. 5.4 Shared memory
switch architecture

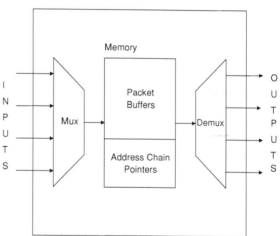

The memory is partitioned by the output ports that packets are addressed to. At one extreme is "full sharing" where the complete memory is shared by all output ports. A problem with full sharing is that a heavily used output port's packets may occupy most or all of the memory, leading to a problem of fairness for other ports. One could implement "complete partitioning" to address the problem. Under complete partitioning, $1/N$ of the memory is dedicated exclusively to each output port. While this solves the original problem, the downside is that when a partition fills up, space may be available in other partitions that can not be used.

Packets from the memory leave the other memory port, are demultiplexed and sent to their respective output ports.

Note that the memory access speed must be at least N times faster than the individual packet arrival rate for each input.

Space Division Switches

Space division switches use a number of parallel paths in the switch to boost throughput (i.e traffic flow). Space division switches are usually built using patterned networks of small switching elements. The fact that the same basic switching element is replicated many times makes this type of design suitable for VLSI implementation. There are many types of space division switches. Some particular examples are crossbars, Banyan networks, hypercubes and the knockout switch [51].

Chapter 6
Multiprotocol Label Switching (MPLS)

6.1 Introduction

The idea behind Multiprotocol Label Switching (MPLS) is simple. In datagram switching each packet is treated as an independent element by routers. If one implements a Differentiated Services architecture, each independent packet is treated according to a policy specifically for its service class. In MPLS, on the other hand, the basic unit is a "flow". Packets belonging to a given source destination flow within an MPLS cloud of a network have a label(s) appended to the packet that is used by routers along a specified path to relay the packet from router to router.

In a sense the MPLS flow is similar to the virtual circuit concept used for technologies such as ATM. The technology of MPLS allows a certain degree of quality of service (QoS) for each flow and simplifies packet forwarding by routers (speeding their operation through the use of simple table lookups based on labels). Virtual private networks with good security can be set up using MPLS. Enhanced traffic engineering and more than one protocol (hence the "M" in MPLS) are also supported by MPLS. For instance, MPLS can carry both IP packets and ATM cells.

Historically, several proprietary antecedents of MPLS were pushed by router vendors (CISCO, IBM and others) in the 1990s. The Internet Engineering Task Force (IETF, see http://www.ietf.org) started working on a standardization effort in 1997 with initial standard release in 2001 [3, 62].

6.2 Technical Details

The technology of MPLS has its own language. A label switched router is a router that follows the MPLS protocols. A set of routers or nodes that are adjacent to each other and form a single administrative domain for routing purposes is an MPLS domain. A router providing connectivity between a domain and nodes outside the

T. Robertazzi, *Basics of Computer Networking*, SpringerBriefs in Electrical and Computer Engineering, DOI: 10.1007/978-1-4614-2104-7_6,
© The authors 2012

Fig. 6.1 An exploded view
of an MPLS label embedded
in an IP packet

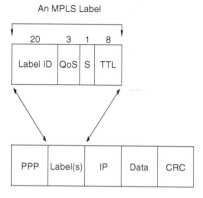

domain is an (MPLS) edge node. A router receiving traffic entering a domain is an
ingress node. Naturally, a router handling traffic leaving a domain is an egress node.

A label switched hop is a single hop connecting two MPLS routers that uses
MPLS labels for forwarding. A label switched path is a sequential path through a
series of label switched routers that packets belonging to a flow transit. This flow is
a forward equivalence class. See Stallings [62] for a more complete listing of MPLS
terminology.

In Fig. 6.1 an MPLS Label is shown both in exploded view and as part of an
datagram packet. In the datagram packet, PPP is a field for a protocol for data link
framing such as with the Point-to-Point protocol [65] and CRC is a cyclic redundancy
code field. The label consists of a 20 bit label number, 3 bit quality of service (class)
designator, a S bit for label stacking (to be discussed) and a time to live field of 8
bits.

Labels are used only locally between a pair of adjacent routers. Thus a router
may see the same label number on a number of incoming links. The label in an
arriving packet is used to do a table lookup to determine the next outgoing line for
the packet as well as the next label to use. A forwarding protocol such as MPLS can
be considered as being between layers 2 (data link) and layer 3 (network).

There are two ways of setting up a label switched path. Under hop-by-hop routing
each label switched router chooses the next hop for a flow. Under explicit routing a
single router such as the ingress or egress router chooses some (loose routing) or all
(strict routing) of the label switched routers a flow traverses [27, 62].

Label switched paths in MPLS can be used as "tunnels". What is tunneling?
Tunneling is a generic concept in networking involving one protocol's packets being
encapsulated as data in another protocol's data field for transmission under the second
protocol. For instance, suppose one has Ethernets in London and Paris. An Ethernet
packet in London (consisting of header and payload) can be placed in the data field
of a SONET frame for transmission to Paris. In Paris the Ethernet packet is removed
from the SONET frame data field and placed on the Parisian Ethernet.

In the MPLS form of tunneling, a packet's path is specified by a label assigned by
a label switched router. Naturally, the label switched path is set up first. Note there is

no need to specify the intermediate routers along the path. That is, the ingress node assigned label allows the packets to follow the (tunneled) path to the egress node.

Multiple labels may be used in the same packet in what is known as label stacking to allow the aggregation of flows. Multiple label switched paths can be aggregated into a single label switched path for some distance in an MPLS network. The labels are used and stacked in LIFO (last in, first out) form in a packet with the top label appearing first. For instance, imagine the use of three levels of labels. The top label could indicate the meta-flow to Europe belonging to some corporation. The middle labels could indicate the corporate traffic destined to particular European countries and the bottom label could indicate the corporate traffic for offices in different cities within each country. The S bit is set to "1" for the bottom label to indicate that it is the last label. One benefit of label stacking is smaller table size in routers.

Let us now take a look at the MPLS label fields. Although the 20 bit "label" field could theoretically support 2^{20} or a bit more than a million labels, actual implementations of label switched routers usually support a much smaller number of labels. Labels can be set up by the Label Distribution Protocol (LDP) or by LDP in conjunction with other protocols [27].

The QoS field with its 3 bits could in theory allow eight classes of traffic. However if one of the bits is used to indicate whether a packet can be discarded in the presence of congestion, only four classes of traffic can be supported with the remaining 2 bits. This is a limitation of MPLS [37].

Finally, the 8 bit time to live (TTL) field plays a similar role to the TTL field in IP packets. Decremented each time a packet makes a hop, the packet is deleted from the network when the TTL field reaches zero to prevent indefinite packet looping in the network.

6.3 Traffic Engineering

Normal routing in datagram-based networks can lead to congestion and poor uti-lization of network resources. In such networks commonly used routing algorithms such as Open Shortest Path First (OSPF) will route packets along the shortest path. This can lead to congestion on shortest paths and under-utilization on longer paths. If there are multiple shortest paths one can use the Equal Cost Multipath (ECMP) option of OSPF [38] but if there is only one shortest path ECMP is not effective. While one might consider manipulation of link costs/metrics this is not practical for large networks [70].

The process of evenly balancing traffic across a network to avoid congestion and under-utilization and to allow the maximum amount of traffic to flow is called traffic engineering. In fact MPLS is well suited to allow traffic engineering because, in terms of congestion mitigation, it is conceptually easier to assign and/or reroute flows than using the indirect method of changing link metrics in a datagram network [29].

Constraint-based routing can be used to computer generate optimal or near-optimal routes using several performance metrics simultaneously. For instance, for purposes of achieving good quality of service one may want to maximize data rate and minimize average link delay and packet loss probability. Algorithms such as shortest path routing use only one metric and are not considered constraint-based routing algorithms. For constraint-based routing individual metrics may be combined to produce a single overall metric, in additive fashion, multiplicative fashion or using a minimum function.

In fact, as Xiao and Ni point out [70], constraint-based routing is a routing methodology and MPLS is a forwarding scheme. It does not matter to MPLS forwarding how routes are chosen. Routing and forwarding are in theory independent. However this is a case where the sum is greater than the parts. The technology of MPLS allows constraint-based routing to use label switched path traffic information on flows through an MPLS domain. The flow paradigm of MPLS is well suited to joint use with constraint-based routing.

6.4 Fault Management

When physical links or routers fail it is desirable to reroute traffic around the fault in a short amount of time so that connections are not lost. Temporally, this can be done in MPLS statically (with preestablished backup paths) or dynamically (as faults occur).

In terms of network topology, there are a variety of ways to reroute traffic with MPLS [6, 33]. Under a "global repair model" an ingress node is made aware of a failure on a path either through a reverse message from the failure site or through a path connectivity test. It then reroutes traffic that was being transported along the original path from the ingress node itself. Under a local repair model, rerouting is done at the point of failure around the fault. Finally under reverse backup repair, traffic flow is reversed at the failure point back to the ingress node where it is continued on alternate path to the destination.

As discussed in Marzo, there are trade offs between the various fault restoration methods. Global repair can be relatively slow, particularly if a path continuity test is used to detect a fault. Reverse backup repair can also be relatively slow. Local repair can be faster. However there are also differences between the schemes in terms of the amount of network resources that need be utilized/reserved.

6.5 GMPLS

At some point it was realized that the MPLS switching paradigm could be carried over to a variety of switching technologies. This extension is called Generalized Multiprotocol Label Switching (GMPLS). The "generalized" in GMPLS comes from the fact that the MPLS protocol is indeed extended to other switching technologies. For instance, for optical (DWDM) networks the concept of a label can be represented by a color (or wavelength) of a signal stream. The development of GMPLS has been aided by groups such as the Internet Engineering Task Force (IETF).

A key advantage of GMPLS is its ability to provide guaranteed QoS and traffic engineering [9]. What GMPLS specifically does is it provides a common control plane for packet (cell) switching, circuit switching, SONET, dense wavelength division multiplexing (DWDM) and fiber switching devices [43, 66]. The architecture of GMPLS provides for protocol capabilities such as signaling (as in RSVP-TE, Resource Reservation Protocol - Traffic Engineering), routing (as in OSPF-TE, Open Shortest Path First-Traffic Engineering), link management (as in LMP, Link Management Protocol) and fault recovery [43].

The traditional hierarchy of IP/ATM/SONET/DWDM is evolving toward an IP/GMPLS/DWDM hierarchy. This will simplify engineering, reduce costs and improve performance [28].

Chapter 7
SONET and WDM

7.1 SONET

Synchronous Optical Networking (SONET) is a popular standard for fiber optic voice and data transmission. It was developed originally by Bellcore, the research and development arm of the local American phone companies in the late 1980s [60]. It was meant to be a standard for fiber optic connections between telephone switches. However, it was a technology at the right place, at the right time and has been extensively used over the years for telephone trunk transmission and internal corporate and governmental traffic. More specifically, it was developed at about the time that there was an interest in providing broadband integrated services digital network (B-ISDN) technology. After its creation it was used to carry ATM traffic, IP packets and Ethernet frames (Wikipedia).

SONET, when it was developed, took into account B-ISDN technology, political and international compatibility concerns. The SONET architecture is elegant and took advantage of LSI and software advances at the time. Development has continued over the years with the introduction of higher and higher standardized data rates.

A typical SONET data rate is abbreviated as STS-n/OC-n where $n = 1, 2, 3, \ldots$ The "STS" indicates the electrical interface and the "OC" indicates the optical interface. The STS-1/OC-1 rate is 51.84 Mbps. Any other STS-n/OC-n rate is n times faster than 51.84 Mbps. For instance, STS-3/OC-3 is at 155.52 Mbps. In fact, STS-3/OC-3 is the lowest SONET rate used in practice. Table 7.1 indicates some of the various SONET rates.

Lower rates, known as virtual tributaries, are also available. For instance, virtual tributary 1.5 (VT1.5) is at 1.728 Mbps. Some virtual tributary rates are indicated in Table 7.2.

Note that VT1.5 is compatible with the T1 rate of 1.544 Mbps and VT2 is compatible with the European version of T1 rate of approximately 2.0 Mbps. SONET is used in the US and Canada. Its cousin, Synchronous Digital Hierarchy (SDH), is used elsewhere and has a greater market share. In fact SONET is considered a version of SDH although it was created first (Wikipedia).

T. Robertazzi, *Basics of Computer Networking*, SpringerBriefs in Electrical and Computer Engineering, DOI: 10.1007/978-1-4614-2104-7_7,

Table 7.1 Some SONET rates

Acronym	Gross rate
STS-1/OC-1	51.84 Mbps
STS-3/OC-3	155.52 Mbps
STS-12/OC-12	622.08 Mbps
STS-48/OC-48	2.48832 Gbps
STS-192/OC-192	9.95328 Gbps
STS-768/OC-768	39.81312 Gbps

Table 7.2 Virtual tributary rates

Acronym	Data rate (Mbps)
VT1.5	1.728
VT2	2.304
VT3	3.456
VT6	6.912

7.1.1 SONET Architecture

SONET data is organized into tables. For STS-1/OC-1, the byte table consists of 9 rows of bytes and 90 columns of bytes (Fig. 7.1). As shown in the figure, the first 3 columns hold frame overhead and the remaining 87 columns hold the payload. Some additional overhead may appear in the payload. Each byte entry in the table holds 8 bits. If digital voice is being carried, the 8 bits represent one voice sample. Uncompressed digital voice consists of 8 thousand samples/sec of 8 bits each (or 64 Kbps). Thus, the SONET STS-1/OC-1 frames are generated at a rate of 8K frames/second.

The protocol layers for SONET go by the names of path, line, section and photonic (see Fig. 7.2 where ATM is being carried over SONET). The functions of the layers are [41]:

- Path = End to end transport as well as functions including multiplexing and scrambling.
- Line = Functions include framing, multiplexing and synchronization.
- Section = Functions include bit timing and coding.
- Photonic = Physical medium.

Overhead of each type appears in a STS-1/OC-1 frame as illustrated in Fig. 7.3. Note that the start of a payload is indicated by a pointer in the line overhead.

The SONET system layers can be viewed in terms of a box type diagram as in Fig. 7.4.

There are two major configurations for the SONET opto-electronic interface at a node. If the fiber starts/ends at the node, one says one has a SONET Add Drop Multiplexer (ADM) in terminal mode. The ADM allows signals to be tapped off or on the fiber. Alternately, one may have fiber passing through the node. That is, a fiber enters from the east, for instance, is converted to an electrical signal, signals

Fig. 7.1 A STS-1/OC-1
SONET frame

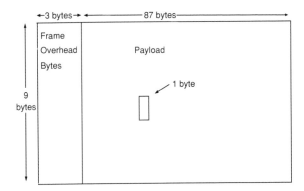

Fig. 7.2 SONET protocol
stack

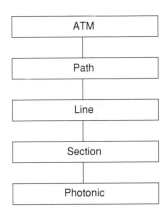

Fig. 7.3 A STS-1/OC-1
SONET frame with overhead
indicated

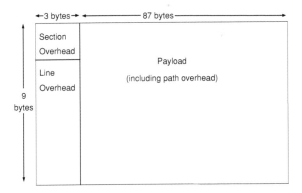

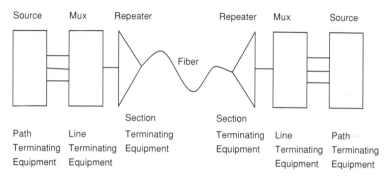

Fig. 7.4 SONET system diagram

Fig. 7.5 A SONET linear
network

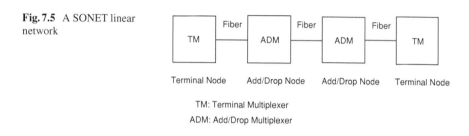

are tapped off and inserted and a new fiber leaves to the west. This is called a SONET
ADM in add/drop mode.

7.1.2 Self-Healing Rings

One of the reasons for the widespread use of SONET is it allows a variety of topologies. The most useful of these are ring topologies. While one can implement linear
add/drop networks (see Fig. 7.5), if a fiber in these is cut or an opto-electronic transceiver fails, one loses complete connectivity.

Typically, rings are implemented with multiple service and protection fibers (see
Fig. 7.6). If a service fiber path fails, a protection (back-up) fiber is switched on its
place. The number of protection fibers can be less than the number of service fibers
if one is willing to live with less redundancy. Also, if all of the fibers between two
adjacent nodes are lost, a sufficient number of protection fiber rerouting can keep a
logical ring in place until repairs are made.

A competitor, at least for data traffic, to SONET is 10 Gbps Ethernet [35].

SONET rates have increased over the years but not by enough for fiber to reach its
full potential unless a second technology, wavelength division multiplexing, is also
used. This WDM technology is discussed in the next section.

Fig. 7.6 A SONET ring
network

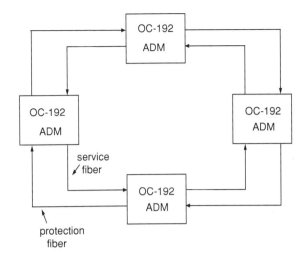

7.2 Wavelength Division Multiplexing (WDM)

In wavelength division multiplexing (WDM), special multiplexing at either end of a fiber can put multiple optical signals in parallel on a single fiber [53]. Thus, instead of carrying one OC-192 signal at about 10 Gbps, a single fiber might carry 40 OC-192 signals or 400 Gbps of traffic. Even terabit range (1000 Gbps) capacity is possible. Current systems can have up to about 160 channels. With each signal at a different optical frequency, WDM is essentially a frequency division multiplexing scheme.

The history of WDM goes back to a fiber glut in the US that existed prior to 1995. After the Bell System divestiture in the 1980s, the competitors in long distance phone service had financial limitations, so relatively low numbers of fiber per path were laid (usually about 16 fibers). But by the end of 1995, the interexchange fibers of the long distance carriers were nearing exhaust. In 1996, 60% of the AT&T network, 84% of the MCI network and 83% of the Sprint network were fully lit [53].

About this time WDM technology became practical. This technology included distributed feedback lasers needed to produce the monochromatic output of WDM and filters to separate signals which are closely packed in frequency and optical amplifiers. In particular, Erbium Dopel Fiber Amplifiers (EDFA) allowed the amplification of optical signals without intermediate electronic conversion.

In 1994, Pirelli/Nortel introduced 4 channel systems and IBM introduced a 20 channel system. Cienna followed with a 16 channel system in 1996. By 97/98, 32 and 40 channel systems were being produced. It should be noted that Cienna's successful WDM products led to a very successful public offering and 196 million dollars in first year revenue (the fastest in corporate history at that point).

Now, using WDM in conjunction with SONET, if one had 32 OC-48 channels, rather than using 32 separate fibers, one could use a 32λ system (32λ means 32 wavelengths) on a single fiber or 8 fibers each with OC-192.

Depending on the wavelength assignment WDM systems may be either conventional/coarse (CWDM) or dense (DWDM). More channels are accommodated with denser channel spacing with DWDM compared to CWDM (Wikipedia).

At first, WDM technology was used in long distance networks but as its costs decreased, metropolitan area network usage followed.

Tunable lasers have been introduced as a way of providing back-up. That is, rather than having 176 fixed wavelength spares for a 176λ system, one tunable laser provides protection against the most likely case, a single bad fixed wavelength laser.

Three generic switching technologies can be used to carry IP traffic over WDM [8]. These are optical circuit switching, packet switching and optical burst switching.

For optical circuit switching to be efficient, such as circuit switching in general, the length of transmissions needs to be significantly greater than the circuit setup time. This is often not true of bursty data traffic. It has also been shown that the circuit establishment problem is in general NP hard. That is, finding an optimal solution is computationally intractable though suboptimal heuristic solution techniques can be used.

The technology for buffering and packet processing, used by packet switching techniques, is not yet mature in the optical area making practical, cost-efficient, systems not possible at this time.

The proponents of optical burst switching feel it has the advantages of both circuit and packet switching. The burst is the fundamental element of optical burst switching. It can be thought of as a variable length packet with a control header and the payload to be carried. An optical burst switching system operates over edge and core routers. Edge (of the network) routers can be either ingress or egress routers. A variable length burst is assembled at an ingress router from multiple IP packets, possibly from multiple hosts. The burst is transported over the core network and its routers. At an egress router the burst is disassembled.

Chapter 8
Grid and Cloud Computing

8.1 Introduction

How does cloud computing differ from grid computing? Grid computing usually involves scheduled scientific or engineering computing done on a distributed network of computers and/or supercomputers. The applications for clouds are often a bit more mundane but also are more encompassing than grids. The basic idea is that organizations can rent time on an outside provider's data center(s) to host applications, software and services that are more traditionally provided by in-house computing facilities. Both grids and cloud computing are accessed through the Internet.

8.2 Grids

A grid is a distributed computing system that allows users to access large amounts of computer power and data storage over a network to run substantial computing tasks. Ian Foster, a leader in grid development, has written (Garritano) that a grid must

- Provide resource coordination without a central control.
- Use standardized and open protocols and interfaces.
- Provide substantial amounts of service involving multiple resource types and non-trivial user needs.

As Schopf points out, the idea of applying multiple distributed resources to work on a computational problem is an old one. It goes back at least to the mid sixties and the "computer utility" paradigm of the MULTICS operating system and work on networked operating systems. Further work on distributed operating systems, heterogeneous computing, parallel distributed computing and metacomputing further explored this area.

Work on the grid concept started in the mid 1990s. Among significant grid features that make it a distinctive problem area are:

T. Robertazzi, *Basics of Computer Networking*, SpringerBriefs in Electrical and Computer Engineering, DOI: 10.1007/978-1-4614-2104-7_8,

- Grids allow site autonomy. Local sites have control over their own resources.
- Grids allow heterogeneity. Grid development work provides standardized interfaces to overcome site diversity.
- Grids are about data (representation, replication, storage in addition to the usual network and computer issues).
- The key person for grids is the user, not the resource owner. Earlier systems sought to boost utilization/throughput for the resource owner. In a grid, machines are selected to fulfill the user's requirements.

A great deal of grid work is related to interfaces and software development to allow different sites to work in concert. However, the grid effort has taken time, enough time that some have questioned its practicality. A balanced discussion of the difficulties and challenges facing grid development appears in Schopf and Nitzberg. Here we mention some of these problems.

- In many cases, grids are being used to solve embarrassingly parallel applications rather than for coordinated distributed computing.
- Users often have to go through a great deal of work to achieve even basic functionality.
- Some science users see making applications suitable for the grid as a "distraction" from getting the science accomplished.
- Funding for adapting applications to a grid environment can be drained, if not blocked, by installation and administration problems.
- System administrators are used to having control over their local resources so grid software can be seen as "threatening."
- Setting up an account in a distributed grid can be complex.

An additional problem is that the number of users has not met the expectations of those who are in favor of grid technology [58].

However the vision of successful grids may make overcoming these difficulties and growing pains worthwhile. Quite a few countries have invested in grid infrastructure to date. Efforts include [58] D-Grid in Germany (http://www.d-grid.de), Grid' 5000 in France, DAS in the Netherlands, PL-Grid in Poland (http://www.plgrid. pl), NAREGI in Japan (http://www.naregi.org) and Open Science Grid (http://www. opensciencegrid.org) and TeraGrid (http://www.teragrid.org) in the United States. There are some grids for communities using large amounts of data such as for particle physics.

The Open Grid Service Architecture is an important grid standard announced in 2002. The service-oriented architecture of OGSA is illustrated in Fig. 8.1.

In Fig. 8.1, note that OGSA services are presented to the application layer. Also, OGSA services make use of web services. Note also this grid architecture has three hardware components: servers, storage and networks. A document discussing the architecture of OGSA can be found on Wikipedia.

The main grid standards setting body is the Open Grid Forum. It was created from the merger of the Global Grid Forum and the Enterprise Grid Alliance in 2006. The Open Grid Forum is responsible for standards such as OGSA, Open Grid Services

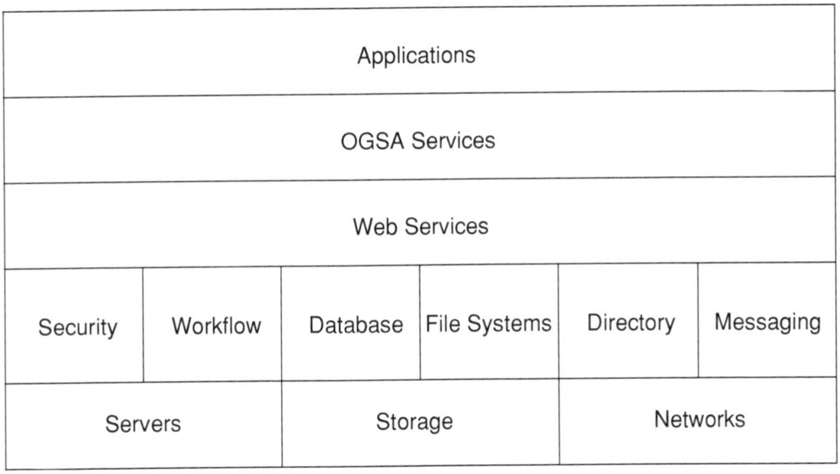

Applications					
OGSA Services					
Web Services					
Security	Workflow	Database	File Systems	Directory	Messaging
Servers		Storage		Networks	

Fig. 8.1 Open Grid Service (OGSA) Architecture

Infrastructure (OGSI), Job Submission Description Language (JSDL) and GridFTP. It has also done recent work on cloud computing. Other organizations who have had grid involvement include the Organization for the Advancement of Structured Information Standards (OASIS), the World Wide Web Consortium, the Distributed Management Task Force, the Web Services Inter-operability Organization and some groups in Internet 2 such as the Peer-to-Peer Working Group, the Middleware Architecture Committee for Education and the Liberty Alliance [5].

For further information readers are referred to Schwiegelshohn [58], http://www.gridforum.org and http://www.globus.org.

8.3 Cloud Computing

Again, under the usual cloud computing paradigm an organization can rent time on a provider's data center(s) resources to host applications, software and services that to date have been provided on in-house computing facilities. There are three main types of cloud computing services [63, 73]:

Software as a Service (Saas): This allows a user (a consumer, student, employee of a business) to use applications accessed through the Internet that are run on the facilities of the entity supplying the service at a data center "inside the cloud." Thus the user only needs a web browser, not the actual application software. The user does not directly manage the cloud infrastructure (servers, storage, etc.).

Infrastructure as a Service (Iaas): Provides a complete computer infrastructure over the Internet (virtual computers, services, storage, etc.).

Platform as a Service (Paas): User essentially has a "virtual platform" accessed through the Internet. The user can load applications using languages and/or tools that are supported by the cloud service provider. Users control applications but not the supporting infrastructure.

Zissis and Lekkas [73] describe four ways in which clouds can be used:

- Private Cloud: a private entity/organization uses cloud facilities dedicated for its own use. These facilities may be administered by the organization itself or a different entity and may be on-site or off-site.
- Community Cloud: the cloud facilities are shared by a community of several related entities. Again, it may be operated by the entities or a different entity and may be on-site or off-site.
- Public Cloud: cloud facilities owned by an entity selling cloud services are offered to the general public, industry, government or a selected group of such.
- Hybrid Cloud: an aggregation of at least two clouds of the previous types that support application and/or data portability for load balancing among the clouds.

8.3.1 Tradeoffs for Cloud Computing

Like any relatively new technology, cloud computing has a number of advantages and disadvantages [32, 73]. Among the advantages are:

- Cost: by "renting" services an organization can avoid large capital investments in infrastructure. Cloud service providers can take advantage of economies of scale through locating data centers on inexpensive real estate and in areas with low power costs and by aggregating the stochastic demand of a number of users into a centralized, expertly managed facility.
- Flexibility: users can quickly have access to computing resources. Provisioning is elastic and fast. Services can be scaled according to client demand.
- New Services: new, innovative applications can be deployed in a straightforward manner. Such applications include parallel batch processing, business analytics and social media such as FaceBook and YouTube.

What of disadvantages? A major issue for organizations thinking about cloud computing is the loss of direct control of their data. Is the entity providing cloud computing services trustworthy? Do they run secure facilities? Are they liable to go out of business suddenly causing a loss of data?

As a niche example, the University of Westminster in the UK transitioned to a Google-based email system. However they continued to maintain records, intellectual property, legal, research and employment information on university owned systems [63].

Related to this is the issue of cloud security. This includes protection of data and integrity of applications. Some feel [42] that we are in a nascent era of cloud attacks by hackers. Cloud computing is built on easy access unlike systems that are protected behind a firewall. Hijacked cloud computing infrastructure could be used for distributed denial of service attacks. Hijacked resources in a cloud could be used for a cross-cloud attacks on cloud users.

Chapter 9
Data Centers

9.1 Introduction

Data centers are networked collections of computers that provide the computational resources for web-hosting, e-commerce, grid and cloud computing and social networking in a centralized location. Generic service platforms for this purpose include Sun's Grid Engine, Google's App engine, Amazon's EC2 platform and Microsoft's Azure platform. In the following we largely follow the excellent treatment in [21].

Virtualization, the ability of a data center to service many independent users while giving each user the impression of a dedicated facility, is important in data center technology for providing good server utilization and for making resource allocation flexible. Because of virtualization, data center management is not a simple problem. The expected trend in data center development is that data centers will become more virtualized, distributed and a "multi-layer infractructure" [21]. This will lead to a number of difficult technological problems.

9.2 Data Centers

9.2.1 Racks

The personal computer-like computers (servers) used in data centers are mounted in racks. A rack will actually contain not only the servers but often storage and specialized devices. A standard rack is 78 inches high, 23–25 inches wide and 26–30 inches deep. Assets mounted in a rack are measured in "U"'s (i.e., 45 mm or about 1.8 inches). A single or dual processor may be 1U. A four socket multiprocessor may be 2U or more. A standard rack can accommodate 42 1U assets [21].

T. Robertazzi, *Basics of Computer Networking*, SpringerBriefs in Electrical and Computer Engineering, DOI: 10.1007/978-1-4614-2104-7_9,

Servers can be in a 13 inch high chassis. Six of these can be inserted into a rack since 6 times 13 is 78 inches (typical chassis height). A chassis will have a power supply, a backplane interconnect, fans and management support. Sixteen 1U servers can be placed in a chassis so that a rack can hold 6 times 16 or 96 servers. "Blades" are modular assets in a chasis.

Racks vary greatly in their complexity. They may or may not include a metal enclosure, rack power distribution, air or liquid cooling at the rack level, a keyboard, video monitor and mouse (e.g., kvm switch) and rack level management unit.

Power is a key consideration in data center design and operation. Kant reports that racks in older data centers may utilize 7 kW per rack and a loaded blade server rack may utilize 21 kW.

9.2.2 Networking Support

Data centers often are equipped with Infiniband, Ethernet, lightweight transport protocols implemented over Ethernet and/or PCI-Express-based backplane interconnect. Wireless networking tends to be a niche technology in data centers. Another issue is that with increasing transmission speeds, protocols need to be lightweight (i.e., simple). However there is a need for some complexity because of security and other needed functionality.

There are at least four types of network access for a data center. This can lead to the use of several (even four) networking technologies in a data center [21].

(1) Client-server network for data center access: This could utilize Ethernet or wireless LAN technology.
(2) Server-server network: This operates at high speed and may utilize Ethernet, Infiniband or other networking technologies.
(3) Access to storage: Historically this may use Fibre channel technology but storage access can also use technologies such as Ethernet or Infiniband.
(4) Management network: This could be Ethernet or a side channel on the main network.

Often the uplinks are "over-subscribed". One speaks of an over-subscription ratio as it may not be possible to achieve the full bisection bandwidth. Bisection bandwidth is the worst case bandwidth between a segmentation of a network into two equal parts. For instance if 15 servers at 1 Gbps each share a 10 GBps link, the over-subscription ratio is 1.5.

Tree networks, traditionally used in data centers, have a problem in that they can have less cross section bandwidth than would be ideal. Fat trees and other types of interconnection networks have been suggested to alleviate this problem.

Mitigating or preventing denial of service attacks is another concern for data centers [21].

9.2.3 Storage

Storage capacity and data generation volume continues to increase at a great rate (some estimates are 50–70% growth a year).

Storage usually has used rotating magnetic technology to date. Because of rotating discs's mechanical construction sequential access is faster than random access. Storage can be responsible for twenty to thirty percent of a data center's power usage. A relatively new technology is solid state storage (non-volatile RAM or NVRAM). Kant [21] suggests that solid state storage will be in a "supporting" role for rotating discs for the near future.

In data centers one has [21]:

- DAS (Direct Access Storage): direct connection to server.
- SAN (Storage Area Network): storage (block-riented) exists across the network.
- NAS (Network Attached Storage): storage (files or objects) exists across the network.

Data intensive applications, streaming and search create large loads on the storage system. It tends to be less expensive to carry all types of networking load on Ethernet but basic Ethernet does not have quality of service (QoS) support.

9.2.4 Electrical and Cooling Support

A medium sized data center can require several megawatts in peak power. Power is provided to the data center on high-voltage lines such three phase 33 kV lines. Transformers on premises step it down to three phase power at 280–480 V. Power then goes to uninterruptible power supplies (UPS). The UPS output is often single phase power at 120–240 V. This is supplied to a power distribution unit which provides electric power to a rack mounted chassis or blade chassis. Power here is stepped down and converted to DC from AC to provide plus or minus 5 V or plus and minus 12 V. Voltage regulators on the mother board change this to even lower voltages for the rails such as 1.1, 5 and 12 V.

With power conversion efficiencies of 85–95% at each step (and 50% or so at the mother board rails) there is a great deal of energy loss and room for improvement. Data center power usage as a percentage of US power consumption doubled from 2000–2007 and is expected to double again by 2012 [21].

Cooling can account for 25% or so of a data center's electric power usage. Air may enter racks at 21°C and leave at 40°C. Cooling necessitates building air conditioning units as well as large chiller plants, air re-circulation systems and fans.

Improving power usage can be done in a number of ways [21]:

- Designing hardware and software for low power consumption.
- Designing the occurrence of hardware power states to minimize power consumption. This includes for CPU's, interconnection networks and for memory.

An example might be a low power sleep state for a piece of hardware when not being actively used.
- In general, power appropriate use and regulation of data center infrastructure.

One metric for measuring the efficiency of a data center is power-usage effectiveness (PUE) [54]. This is the ratio of the total power used by a data center to the power consumed by just the computers and networking equipment. Schneider and Hardy report that Google achieved an "overall" PUE of 1.13 at the end of 2010 and Facebook "believes" its new data centers will have a PUE value of 1.07 or lower.

9.2.5 Management Support

There is management controller within each server known as the baseband management controller (BMC). Functions of the BMC include [21]:

- Booting up and shutting down the server.
- Managing hardware and software alerts.
- Monitoring hardware sensors.
- Storing configuration data in devices and drivers.
- Remote management capabilities.

9.2.6 Ownership

To date data centers tend to have a single owner. However virtual data centers, where an organization may have ownership rights for the data center but not for the physical manifestation of the data center are a possibility. A virtual data center would be based on longer term agreements than cloud computing.

9.2.7 Security

As data centers increase in size, there is a need for scalable security solutions. One difficulty is that an attack may originate from an organization sharing a data center.

Chapter 10
Advanced Encryption Standard (AES)

10.1 Introduction

The Advanced Encryption Standard (AES) is the encryption standard approved in 2000 under the auspices of the US government, originally for civilian cryptographic use. In 2003 the United States government approved the use of AES for classified and secret information. Versions of AES for 192 bit and 256 bit keys are approved for top secret information. The National Institute of Standards and Technology, which led the 3 year approval process for AES, has certified in excess of 1000 products incorporating AES [10]. The Advanced Encryption Standard is incorporated in standards, algorithms and requests for comments from IEEE (Institute of Electrical and Electronics Engineers), IETF (Internet Engineering Task Force), ISO (International Organization for Standardization)and 3GPP (Third Generation Partnership Project, see wireless chapter). The use of AES in hardware and smart cards is taking longer but progress is being made.

10.2 DES

The predecessor to AES was DES (Digital Encryption Standard). The National Bureau of Standards, which eventually became NIST, called for proposals for a block encryption standard (i.e. encrypting a block of data at a time rather than a continuous stream of data as a stream cipher would) in 1973 [7]. The only practical candidate was one from IBM. This was modified into what became known as DES.

There were some public issues concerning DES. The original 128 bit key was reduced to 56 bits in DES. Some changes were also made to scrambling boxes known as "S boxes". Some felt that without knowing the reasons for the DES modifications it was difficult to make an assessment of its security. When differential cryptanalysis[1]

---

[1] A technique where small changes to input are correlated with output changes to attempt to find the key.

T. Robertazzi, *Basics of Computer Networking*, SpringerBriefs in Electrical and Computer Engineering, DOI: 10.1007/978-1-4614-2104-7_10,

was discovered some time later it was found that DES was more resistant to differential cryptanalysis because of the S box changes. It was conjectured that IBM and NSA (the US code breaking and creation agency) knew of differential cryptanalysis and the changes were made for this reason [7].

As Burr [7] puts it, DES is the standard against which all block ciphers are compared.

In 2004 NIST withdrew DES through a version using three keys/steps of encryption and decryption, triple DES, is still approved.

10.3 Choosing AES

It is interesting to briefly discuss the selection process that led to the choice of the current AES (the summary here is largely based on the excellent discussion in Borr [7]).

At the time of the selection of DES, the thought was that encryption is best done in hardware. But DES is not well suited for software operation. One needs to shuffle/scramble 4 or 6 bits. This was fast in hardware but slow on computers. Moreover triple DES is three times slower than DES. As the years went by, software encryption became more important.

NIST created a selection process for AES for federal and international business purposes. Among the properties an acceptable AES would need are [7]:

- Should be a block cipher.
- It should be at least as secure as triple DES.
- 128 bit block size.
- Options for key sizes of 128, 192 and 256 bits.
- It should be unclassified and open to the public (not patented and royalty free).

Fifteen qualifying groups submitted proposals in 1998. The next year there were five final candidates:

- *MARS* from IBM (US).
- *RC6* from RSA Data Systems (US).
- *Rijndael* from Joan Daemen and Vincent Rijmen (Belgium).
- *Serpent* from Ross Anderson, Eli Biham and Lars Knudsen (UK, Israel and Denmark).
- *Twofish* from a team of American companies and academics (US).

Burr's 2003 article describes each contender.

In 2000 NIST selected Rijndael as the new AES. Rijndael was the most popular finalist in polls at the most recent AES conference. The international cryptographic community was well disposed towards it. Finally, the selection of a non-US cipher made international acceptance smoother. Rijndael was made the official AES in Dec. 2001.

10.4 AES Issues

NIST's criteria for selecting AES included security, performance and intellectual property.

10.4.1 Security Aspect

It was thought about 2003 [7] that an 80 bit key would provide sufficient protection against exhaustive search for a limited number of years. Moreover if Moore's law (computing power doubling every 18–24 months) remains true a 128 bit key would provide protection until 2066 and a 256 bit key would provide protection for several centuries. Of course what form technology will take centuries from now is hard to predict and breakthroughs such as practical and robust quantum computing may change the situation radically.

Most symmetric key algorithms consist of multiple iterations/repetitions of a (scrambling) function called a round function [7]. People evaluating codes will try to find short cuts on simpler versions of an encryption algorithm that use less than the full set of rounds. The attack will then be tried on the complete algorithm. People creating encryption algorithms estimate the number of rounds needed for security and add to this additional rounds as a safety margin.

An encryption algorithm is deemed secure if no short cuts are known. The longer it has been evaluated the more confidence one has in its security. But encryption products are different from typical products. With a typical product it is designed, tested and fielded. While surprises are certainly possible after it is fielded, they are rare. With an encryption algorithm no matter how much effort goes into its design, one is never quite sure in the back of one's mind if there is some clever way to crack it.

For the full versions of the final contenders in the AES competition no shortcuts were found. Attacks were found for versions of the final contenders with fewer rounds. Actually it was not possible to differentiate between the contenders on the basis of security. As reported by Burr [7] the best attack against Rijndael worked on 7 of 10 rounds and needed 2^{127} pairs, 2^{101} bytes of memory and 2^{120} operations. Subsequent attacks aimed at breaking AES have only been successful on reduced versions of AES [10].

There are arguments about whether a simple or a complex encryption algorithm provides better security. A simpler algorithm can be better understood but some feel because it is simple a new attack may crack (break) it. For a complex algorithm if one part is broken the other parts may still provide backup security, some hold. But a complex algorithm may be difficult to fully analyze so there may be an unknown way an attacker may succeed in breaking it. The creators of Rijndael believe simplicity is the better virtue [10].

10.4.2 Performance Aspect

The performance of the candidate algorithms in the AES selection process was first discussed for 32 bit Pentiums. There was at least moderate performance for all of the contenders on Pentiums. All of the contenders had a better performance than triple DES. Performance was also examined based on RISC processors, Itaniums, 8 bit embedded microprocessors, digital signal processors (DSP), field programmable gate arrays (FPGA) and application-specific integrated circuits (ASIC) [7].

10.4.3 Intellectual Property Aspect

It was desired that AES be available royalty free on world wide basis. This goal was made easier because DES patents had expired and there were quite a few algorithms available that were not patented. It was found by NIST that the final contenders' algorithms did not infringe on any existing patent.

There was also some question as to whether NIST should choose one winner or more than one. Straw polls at the third AES conference indicated that one winner was the community's preference. One reason [7] is that choosing several algorithms might lead to compatibility problems, require more chip real estate and there would be a higher probability of intellectual property issues.

10.4.4 Some Other Aspects

It takes more than encryption algorithms to make secure systems. For instance, sending an email may require a public key transport algorithm, public key signature algorithm, a hash algorithm and then the (AES) encryption algorithm [7]. Work has gone on in standardizing other strong algorithms in addition to encryption algorithms.

Another issue is that block ciphers such as DES and AES can be used in a number of well-known "modes" of operation. These include:

- *Electronic Code Book Mode*: A plaintext number of bits is mapped into the same number of encrypted bits. A weakness is that an attacker may swap sections of the code with code of their own.
- *Cipher Block Chaining Mode*: One linearly chains block ciphers, so replacing a block causes unintelligible code after that point.
- *Cipher Feedback Mode*: Uses feedback in the coding process for byte-by-byte encryption.
- *Stream Cipher Mode*: A simple way to convert any block cipher to a stream cipher (i.e. encrypting a continuous stream of data rather than a block at a time).

A weakness of chained and feedback modes is that they cannot be parallelized and they cannot be arbitrarily accessed (one needs to start at the beginning). To address this a counter mode was introduced which is parallelizable [7].

Other modes have been proposed to NIST. Some are parallelizable and do encryption, authentication and integrity protection for just a bit more than the cost of encryption. However the inventors filed patents on these, so there are some intellectual property issues.

Bibliography

1. S. Ahmadi, An overview of next-generation mobile WiMAX technology. IEEE Commun. Mag. **47**(6), 84–98 (2009)
2. I.F. Akyildiz, D.M. Gutierrez-Estevez, E.C. Reyes, The evolution to 4G cellular systems: LTE advanced. Phys. Commun. **3**, 217–244 (2010)
3. G. Armitage, MPLS: the magic behind the myths. IEEE Commun. Mag. **38**, 124–131 (2000)
4. A. Bacioccola, C. Cicconetti, C. Eklund, L. Lenzini, Z. Li, E. Mingozzi, IEEE 802.16: history, status and future trends. Comput. Commun. **33**, 113–123 (2010)
5. M. Baker, A. Apon, C. Feiner, J. Brown, Emerging grid standards. Computer **38**(4), 43–50 (2005)
6. O. Banimelhem, J.W. Atwood, A. Agarwal, Resiliency issues issues in MPLS networks. Canadian conference on electrical and computer engineering, CCGEI 2003, Montreal, Quebec, 1039–1042 May 2003
7. W.E. Burr, Selecting the advanced encryption standard. IEEE Secur. Priv. **1**(2), 43–52 (2003)
8. P.K. Chandra, A.K. Turuk, B. Sahoo, Survey on Optical Burst Switching in WDM Networks. in *Proceedings of the Fourth International Conference on Industrial and Information Systems, ICIIS 2009*, pp. 83–88 (2009)
9. J.M. Chung, H.K. Khan, H.M. Soo, J.S. Reyes, G.Y. Cho, Analysis of GMPLS Architectures, Topologies and Algorithms. in *Proceedings of the 2002 45th Midwestern Symposium on Circuits and Systems MWSCAS-2002* vol. 3 (2002), pp. 284–287
10. J. Daemen, V. Rijmen, The first 10 years of advanced encryption. IEEE Secur. Priv. **8**(6), 72–74 (2010)
11. J. D'Ambrosia, D. Law, M. Nowell, 40 Gigabit Ethernet and 100 Gigabit Ethernet Technology Overview. *Ethernet alliance*, http://www.ethernetalliance.org (2008)
12. J.P. Dunning, Taming the blue beast: a survey of bluetooth-based threats. IEEE Secur. Priv. **8**(2), 20–27 (2010)
13. C. Eklund, R.B. Marks, K.L. Stanwood, S. Wang, IEEE Standard 802.16: a technical overview of the wireless MAN interface for broadband wireless access. IEEE Commun. Mag. 98–107 (2002)
14. T. Garritano, Globus: an infrastructure for resource sharing. Clusterworld **1**(1), 30–31 50
15. R.C. Garroppo, S. Giordano, L. Tavanti, Implementation frameworks for IEEE 802.11s systems. Comput. Commun. **33**, 336–349 (2010)
16. L. Goldberg, 802.16 wireless LANs: a blueprint for the future? Electron. Des. **4**, 44–52 (1997)
17. P. Grun, Introduction to infiniband for end users. *Infiniband Trade Association*, http://www.infinibandta.org (2010)

18. J.C. Haartsen, S. Mattisson, Bluetooth—a new low-power radio interface providing short-range connectivity. Proc. IEEE **88**(10), 1651–1661 (2000)
19. G.R. Hiertz, D. Denteneer, S. Max, et al., IEEE 802.11s: the WLAN mesh standard. IEEE Wireless Commun., Feb. 104–111 (2010)
20. J.M. Kahn, R.H. Katz, K.S.J. Pister, Emerging challenges: mobile networking for smart dust. J. Commun. Netw. **2**(3) (2000)
21. K. Kant, Data center evolution: a tutorial on state on the art, issues and challenges. Comput. Netw. **53**, 2939–2965 (2009)
22. S. Kapp, 802.11: leaving the wire behind. IEEE Internet Comput. **6**(1), 82–85 (2002)
23. M.S. Kuran, Y. Tugcu, A survey on emerging broadband wireless access technologies. Comput. Netw. **51**, 3013–3046 (2007)
24. K.S. Kwak, S. Ullah, N. Ullah, An overview of IEEE 802.15.6 standard. in *3rd International Symposium on Applied Sciences in Biomedical and Communication Technologies (ISABEL)*, pp. 1–6 (2010)
25. R.O. LaMaire, A. Krishnan, P. Bhagwat, J. Panian, Wireless LANs and mobile networking standards and future directions. IEEE Commun. Mag. **34**(8), 86–94 (1996)
26. B. Latré, B. Braem, I. Moerman, C. Blondia, P. Demeester, A survey on wireless body area networks. Wireless Netw. **17**, 1–18 (2011)
27. J. Lawrence, Designing multiprotocol label switching networks. IEEE Commun. Mag. 134–142 (2001)
28. A. Leon-Garcia, L.G. Mason, Virtual network resource management for next-generation networks. IEEE Commun. Mag. 102–109 (2003)
29. T. Li, MPLS and the evolving internet architecture. IEEE Commun. Mag. 38–41 (1999)
30. van der E. Linde, G.P. Hancke, An investigation of bluetooth mergence with ultra wideband. Ad Hoc Netw. **9**(5), 852–863 (2011)
31. M. Littlewood, I.D. Gallagher, Evolution toward an ATD multi-service network. British Telecom Technol. J. **5**(2) (1987)
32. S. Marston, Z. Li, S. Bandyopadhyay, J. Zhang, A. Ghalsasi, Cloud computing—the business perspective. Decis. Support Syst. **51**, 176–189 (2011)
33. J.L. Marzo, E. Calle, C. Scoglio, T. Anjali, QoS online routing and MPLS multilevel protection: a survey. IEEE Commun Mag. 126–132 (2003)
34. M. Mauve, H. Hastenstein, A. Widmer, A survey of position-based routing in mobile ad hoc networks. IEEE Netw. **15**(3), 30–39 (2001)
35. C. Meirosu, P. Golonka, A. Hirstius et al., Native 10 Gigabit ethernet experiments over long distances. Future Gener. Comput. Syst. **21**, 457–468 (2005)
36. R.M. Metcalfe, D.R. Boggs, Ethernet: distributed packet switching for local computer networks. Commun. ACM **19**, 395–404 (1976)
37. C. Metz, C. Barth, C. Filsfils, Beyond MPLS...less is more. IEEE Internet Comput. 72–76 (2007)
38. J. Moy, OSPF Version 2. RFC 2178 (1998)
39. C.S.R. Murthy, B.S. Manoj, Ad Hoc Wireless Networks: Architectures and Protocols (Prentice-Hall, Upper Saddle River, 2004)
40. M. Nowell, V. Vusirikala, R. Hays, Overview of Requirements and Applications for 40 Gigabit and 100 Gigabit Ethernet. *Ethernet alliance*, version 1.0, http://www.ethernetalliance.org (2007)
41. R.O. Onvural, *Asynchronous Transfer Mode Networks: Performance Issues*, 2nd edn. (Artech House, 1995)
42. R.M. Pacella, Hacking the Cloud. *Popular Science*, (2011) 68–72
43. D. Papadimitriou, D. Verchere, GMPLS user-network interface in support of end-to-end rerouting. IEEE Commun. Mag. 35–43 (2005)
44. I. Papapanagiotou, D. Toumpakaris, J. Lee, M. Devetsikiotis, A survey on next generation mobile WiMAX networks: objectives, features and technical challenges. IEEE Commun. Surv. Tutor. **11**(4), 3–18 (2009)

45. C.E. Perkins, *Ad Hoc Networks*. (Addison-Wesley, Boston, 2001)
46. S.W. Peters, R.W. Heath, Jr., The future of WiMAX: multihop relaying with IEEE 802.16j. IEEE Commun. Mag. 104–111 (2009)
47. I. Poole, What Exactly is 802.11n? Commun. Eng. 46–47 (2007)
48. I. Poole, What Exactly Is... LTE? Commun. Eng. 46–47 (2007)
49. J.M. Rabaey, M.J. Ammer, J.L. daSilva Jr. et al., PicoRadio supports ad hoc ultra-low power wireless networking. Computer **33**(7), 42–48 (2000)
50. M. Rinne, O. Tirkkonen, LTE, the radio technology path towards 4G. Comput. Commun. **33**, 1894–1906 (2010)
51. T.G. Robertazzi, in *Peformance Evaluation of High Speed Switching Fabrics and Networks: ATM, Broadband ISDN and MAN Technology*, IEEE Press (now distributed by Wiley) (1993)
52. T.G. Robertazzi, *Networks and Grids: Technology and Theory* (Springer, New York, 2007)
53. J.P. Ryan, WDM: north american deployment trends. IEEE Commun. Mag. **36**(2), 40–44 (1998)
54. D. Schneider, Q. Hardy, Under the hood at google and facebook. IEEE Spectr. **48**(6), 63–67 (2011)
55. J.M. Schopf, B. Nitzberg, Grids: the top ten questions. Sci. Program., IOS Press, **10**(2), 103–111 (2002)
56. M. Schwartz, Telecommunication Networks: Protocols, Modeling and Analysis. (Addison-Wesley, Reading, 1987)
57. L. Schwiebert, S.K.S. Gupta, J. Weinmann, Research challenges in wireless networks of biomedical sensors. ACM Sigmobile 151–165 (2001)
58. U. Schwiegelshohn, R.M. Badia, et al, Perspectives on Grid Computing. Futur. Gener. Comput. Syst. **26**, 1104–1115 (2010)
59. R.C. Shah and J.M. Rabaey, Energy Aware Routing for Low Energy Ad Hoc Networks. in *Proceedings of the 3rd IEEE Wireless, Communications and Networking Conference*, pp. 350–355 (2002)
60. C.E. Siller (ed.), *SONET/SDH: A Sourcebook of Synchronous Networking* (Wiley-IEEE Press, 1996)
61. S. Siwamogsatham, 10 Gbps Ethernet. http://www.cse.wustl.edu/jain. (1999)
62. W. Stallings, *High-Speed Networks and Internets: Performance and Quality of Service* (Prentice-Hall 2002)
63. N. Sultan, Cloud computing for education: a new dawn. Int. J. Inf. Manag. **30**, 109–116 (2010)
64. A. Tanenbaum, *Computer Networks* 3rd ed. (Prentice-Hall 1996)
65. A. Tanenbaum, *Computer Networks*, 4th ed. (Prentice-Hall 2003)
66. M. Tatipamula, F. Le Faucheur, T. Otani, H. Esaki, Implementation of IPv6 services over a GMPLS-based IP/optical network. IEEE Commun. Mag. 114–122 (2005)
67. S.J. Vaughan-Nichols, Will 10-Gigabit ethernet have a bright future? Computer 22–24 (2002)
68. S.J. Vaughan-Nichols, Will the new Wi-Fi fly? Computer 16–18 (2006)
69. http://www.wikipedia.org
70. X. Xiao, L.M. Ni, Internet QoS: A big picture. IEEE Netw. 8–18 (1999)
71. D. Yang, Y. Xu, M. Gidlund, Coexistence of IEEE802.15.4 based networks: a survey. in *Proceedings of the 36th Annual IEEE Conference on Industrial Electronics (IECON 2010)*, 201. pp. 2107–2113
72. J. Zheng, M.J. Lee, Will 802.15.4 make ubiquitous networking a reality? A discussion on a potential low power, low bit rate standard. IEEE Commun. Mag. 140–146 (2004)
73. D. Zissis, D. Lekkas, Addressing cloud computing security issues. Futur. Gen. Comput. Syst. article in press as of April (2011)

Index

T. Robertazzi, *Basics of Computer Networking*, SpringerBriefs in Electrical
and Computer Engineering, DOI: 10.1007/978-1-4614-2104-7,
© The authors 2012